T0073825

GETTING STARTED ON TIME-RESOLVED MOLECULAR SPECTROSCOPY

Getting Started on Time-Resolved Molecular Spectroscopy

Jeffrey A. Cina

University of Oregon

OXFORD
UNIVERSITY PRESS

Great Clarendon Street, Oxford, OX2 6DP,
United Kingdom

Oxford University Press is a department of the University of Oxford.
It furthers the University's objective of excellence in research, scholarship,
and education by publishing worldwide. Oxford is a registered trade mark of
Oxford University Press in the UK and in certain other countries

Impression: 1

Published in the United States of America by Oxford University Press
198 Madison Avenue, New York, NY 10016, United States of America

British Library Cataloguing in Publication Data

Data available

Library of Congress Control Number: 2021945727

ISBN 978–0–19–959031–5

DOI: 10.1093/oso/9780199590315.001.0001

Printed and bound by
CPI Group (UK) Ltd, Croydon, CR0 4YY

For Barbara, of course.

Preface

Well, it has been quite an adventure so far. It was my good fortune to be drawn into the theory of ultrafast spectroscopy early in the femtosecond era, due in large part to the prior influence of Rick Heller's wave-packet descriptions of continuous-wave spectroscopies (thank you, Laurie!); Bob Silbey's coaching during the 1980's, when multi-pulse optical-phase-controlled *picosecond* measurements were first under consideration; and the pleasure and long-term benefit of collaboration with Norbert Scherer, Stuart Rice, Graham Fleming, and other co-workers at Chicago. At that stage, one had to find one's own way, and for me that way started with ordinary time-dependent perturbation theory as I'd learned it in Bob Harris's (personally life-changing) quantum mechanics classes at Berkeley, coupled with a desire to illuminate the dynamics underlying optical measurements in terms of the evolving nuclear wave packets that accompany each molecular electronic state.

It hadn't occurred to me to write this book, or any other, but in 2011 I was in a bit of a hiatus that came at the end of a multi-year group-reading project with co-workers at Oregon on L&L's *Electrodynamics of Continuous Media.* At Bob Mazo's suggestion, Sonke Adlung from Oxford University Press called sometime that year and asked if I might like to write a book. It happened that I was running a fever at the time and said, "Sure!"

So, here you have it, the result of the succeeding decade of puzzlement, formulation, and reformulation—aided by the patience and helpful advice of the numerous collaborators and colleagues whose names are among those listed in my Acknowledgements, hopefully without any inadvertent omissions.

There are valuable treatises already available on the principles of nonlinear optical spectroscopy of molecular and material systems, notably those by Shaul Mukamel and Minhaeng Cho.[1,2,3,4] The coverage of those works goes beyond what is presented here. So why write another? I believe that the ultrafast community could benefit from a deliberately accessible, stepwise treatment (which doesn't mean an easy one), more of a textbook than a comprehensive exposition, which can serve as a bridge between the graduate-level training in quantum mechanics that's standard in Physical Chemistry programs and the advanced formulations that serve as guidebooks for those practicing in the field. The standard approach is to start from the equilibrium density

[1]S. Mukamel, *Principles of Nonlinear Optical Spectroscopy* (Oxford University Press, New York, 1999).

[2]M. Cho, *Two-Dimensional Optical Spectroscopy* (CRC Press, Boca Raton, 2009).

[3]D. J. Tannor, *Introduction to Quantum Mechanics. A Time-Dependent Perspective* (University Science Books, Sausalito, 2007).

[4]J. Yuen-Zhou, J. J. Krich, I. Kassal, A. Johnson, and A. Aspuru-Guzik, *Ultrafast Spectroscopy: Quantum Information and Wavepackets*, (IOP Publishing, Bristol, 2014).

matrix of the target molecule and express nonlinear optical signals as a convolution of the appropriate nonlinear optical response functions with the electric field of the incident laser pulses. In addition, informed by very early analyses of multi-wave mixing in nonlinear optical crystals, conventional descriptions are often couched in terms of optical wave propagation in extended media. Everything in the present text is (or should be!) physically consistent with the widely applied existing theoretical analyses.

Applications of existing descriptions sometimes tend, in my view, to lose track of the molecular-level dynamics underlying ultrafast signals, in part because those descriptions compel, or at least encourage one to think about both bra- and ket-sides of the density operator simultaneously. On the other hand, informed by the superposition principle, framing things at least initially as a sum of terms in Hilbert space having various orders in the external-field strengths makes it easier to think about the pertinent contributions to the molecular state one at a time. What is developed, analyzed, and interpreted in terms of Hilbert-space wave functions can easily be converted to a density-matrix description through a manipulation that takes just a couple of lines.

The wave-propagation picture of nonlinear optical response, while important for experimental purposes, tends to obscure the fact that, even in an extended sample, each molecule can often be regarded as undergoing absorption, fluorescence, and Raman scattering all by itself, as well as acquiring the nonlinear induced dipole moments that give rise to time-resolved signals. One goal of this text is to formulate as much as possible in terms of nonlinear induced molecular dipoles—these are ultimately to be expressed as quantum mechanical overlaps between pairs of multi-pulse wave packets—and to derive any necessary macroscopic wave-propagation aspects, such as wave-vector matching, from that microscopic starting point.

The strength of treatments based on nonlinear optical response functions is their generality. Nonlinear optical signals are to be derived from that fundamental underpinning by convolving the relevant response functions with the actual form of the incident laser pulses by multidimensional integration over time variables. It is the case, though, that such an approach encodes much quantum dynamical information that is not of immediate use in calculating or interpreting the actual laser-driven dynamics under consideration. Those strategies sometimes tend, in practice, to invite simplified comprehensive but phenomenological descriptions that do not attach directly to the specific, perhaps non-generic form of the regions of the molecular potential energy surfaces that govern the measured signal.

The treatment put forward here reverses the order of operation between quantum mechanical averaging and integration over time. It carries out the latter first, with the use of *pulse propagators*, operators analogous to those sometimes used in magnetic resonance spectroscopy, which encapsulate the influence of a nonzero-duration laser pulse within an instantaneously acting quantum mechanical operator. The reduced pulse propagators that transfer nuclear wave packets between electronic potential energy surfaces under the influence of the external fields, reshaping them in the process, capture the elements of coherent control that are inherent in short-pulse optical spectroscopy. This approach puts the focus on the motion of nuclear wave packets in the optically accessed regions of the electronic potential energy surfaces on which they evolve. It declines the frequent practice in work-ups based on nonlinear response functions of idealizing the characteristics of the experimentally available light pulses and, in effect, regarding those response functions themselves as constituting signals, rather than signal transducers. As mentioned above, the working expressions arrived at here present signal contributions as overlaps between well characterized individual multipulse nuclear wave packets.

For ease of use and at the cost of some redundancy, I've tried to make each chapter of this book usable and comprehensible on its own. There are some slight inconsistencies of notation between chapters, so unless the reader is advised to do so, applying a formula from one chapter to an equation in another should be done with caution.

In lieu of end-of-chapter problems, there are many boxed exercises embedded in the text. By and large, these represent the type of derivation or physical analysis that I work through to consolidate my own grasp of the ideas at hand. It's my guess that readers who excuse themselves from these exercises will be sacrificing something in the depth of their understanding.

The final portion of most chapters consists of illustrative signal calculations. Carrying these out or interpreting their form is both our reward for slogging through the underlying theory and a pale substitute for the experiential satisfaction of performing actual measurements. The molecular models that are targeted in these calculations are chosen to facilitate a thorough interpretation in terms of the underlying molecular dynamics. It is hoped that these illustrations will help inform what may be the less complete interpretations that are possible with more complicated experimental targets.

There is a smattering of references throughout the text but no comprehensive bibliography. The works cited are just a few of those I found helpful or inspiring. Many others could undoubtedly have served as well or better; the ones I mention can perhaps serve as points of entry in the search for other relevant examples.

With help from the many colleagues who have read portions of this text, I've tried hard to root out conceptual misdirection and outright errors. From experience, I know that there's no such thing as a *small* mistake in a scientific text. My sincere apology for any that may remain.

Personal taste has a significant influence in science (gasp!), and this textbook reflects my own. I hope, though, that the treatment given here will be to the liking of at least *some* readers. I presume further to hope that, after working their way through this book, those who persevere in doing so will be equipped to make the best use of more sophisticated methodologies, gaining as much physical insight as possible and avoiding some of the pitfalls to which applications of those approaches occasionally give rise.

Jeff Cina
University of Oregon
2021

Acknowledgements

My heartfelt thanks to the following individuals and organizations: y'all know what for. John Adamovics, Paul Alivisatos, Rise Ando, Ara Apkarian, Sonke Adlung, Natasha Aristov, Alan Aspuru-Guzik, Bridgette Barry, Mark Berg, Berkeley Chemists for Peace, Jason Biggs, Sandra Bigtree, Isabella Bischel, Eric Bittner, Harry Bonham, Stephen Bradforth, Bill Braunlin, Paul Brumer, Irene Burghardt, Carlos Bustamante, Laurie Butler, Carl Bybee, Jianshu Cao, Roger Carlson, Howard Carmichael, Sister Carola FSPA, Craig Chapman, Xiaolu Cheng, Minhaeng Cho, Gerri Hoffman Cina, Merrill T. Cina, Zoë Cina-Sklar, Rob Coalson, Devin Daniels, Joshua Daniels, Julia Daniels, Rebecca Daniels, Peter Dardi, Brent Davidson, Jahan Dawlaty, Peter Dewey, Jenni Dobbins, Mike and Carol Drake, Camille and Henry Dreyfus Foundation, Tom Dyke, Ellen Epstein, my many friends at Espresso Roma, Elaine Finley Giannone, Graham Fleming, Joan Florsheim, Ignacio Franco, David Frank, Karl Freed, Sarah Fuchs, Rosemary Garrison, Ibrahim Gassama, Diane Gerth, Michelle Golden, Evi Goldfield, Eoghan Gormley, John Simon Guggenheim Memorial Foundation, John Hardwick, Alex Harris, Charles Harris, Christine Harris, Jerry Harris, Katherine Harris, Robert A. Harris (my former PhD advisor, an exemplar of science, and a lifelong friend), Ray Heath, Rick Heller, Eric Hiller, Maddie Holst, Travis Humble, Katharine Hunt, Heide Ibrahim, Deb Jackson, Suggy Jang, Truus Jansen, David Jonas, Tara Jones, Taiha Joo, Chanelle Jumper, Joe Kao, Mike Kellman, Alexis Kiessling, Dmitri Kilin, Young-Kee Kim, Phil Kovac, Peter Kovach, Cindy Larson, Larry LeSueur, Sister Leora FSPA, Susan Levine-Friedman, Mark Limont, Katja Lindenberg, Giulia Lipparini, Andy Marcus, Alex Matro, Karin Matsumoto, Jack Maurer, Bob Mazo, David McCamant, Alden Mead, Horia Metiu, Stan Micklavzina, Coleen Miller, Florabelle Moses, Shaul Mukamel, Sid Nagel, US National Science Foundation, Keith Nelson, Kenji Ohmori, Takeshi Oka, Janine O'Guinn, Colleen O'Leary, April Oleson, Mrs. Ruth Patterson, Barbara Perry, Fred Perry, David Picconi, David Pratt, Jim Prell, Mike Raymer, Tom Record, Stuart Rice, Philip Richardson, Mary Rohrdanz, Victor Romero-Rochin, Sandy Rosenthal, Sister Rosilda FSPA, Penny Salus, Marsha Saxton, Rich Saykally, Norbert Scherer, Greg Scholes, Moshe Shapiro, Mike Sheahan, Yu-Chen Shen, Nancy Shows, Robert J. Silbey (my postdoctoral advisor, a gem of a human being), Michael Sipe, **Barbara Sklar**, Martin Smith, Tim Smith, Peter Straton, Joe Subotnik, Susie Cina Sullivan, David Tannor, Judithe Thompson, Nacho Tinoco, Dr. Brandy Todd, Emma Tran, Daniel Turner, Rob Tycko, David Tyler, Lowell Ungar, my fellow members of Oregon's faculty labor union United Academics, Marian Valentine, Steven van Enk, Hailin Wang, John Waugh, Bob Weiss, Julia Widom, Kevin Wiles, Cathy Wong, Claude Woods, Duane C. Wrenn and Leticia Steuer of Energetic Soul, John Wright, Joel Yuen-Zhou, Larry Ziegler.

Contents

1	**Short-pulse electronic absorption**	**1**
	1.1 Basic set-up	1
	1.2 Energy changes of molecule and field	2
	1.3 Expectation values	3
	1.4 The Heller formula	5
	1.5 Systems starting in thermal equilibrium	6
	1.6 Example absorption calculations	7
2	**Adiabatic approximation**	**11**
	2.1 Molecular Hamiltonian	11
	2.2 Molecular eigenstates	12
3	**Transient-absorption spectroscopy: Making ultrashort pulses worthwhile**	**17**
	3.1 Model Hamiltonian and signal expression	17
	3.2 Transient-absorption dipole	20
	3.3 Exemplary calculations	23
4	**How fissors works: Femtosecond stimulated Raman spectroscopy as a probe of conformational change**	**34**
	4.1 Basic idea	34
	4.2 Signal formation	36
	4.3 Fissors dipole	38
	4.4 Example signal calculation	45
5	**Transient-absorption reprise: Taking advantage of vibrational adiabaticity**	**53**
	5.1 Transient-absorption signal under vibrational adiabaticity	53
	5.2 Calculated transient-absorption signals	55
6	**Two and a half approaches to two-dimensional wave-packet interferometry**	**61**
	6.1 Introduction	61
	6.2 Measured quantities	62
	6.3 Quantum mechanical aspects	76
	6.4 Example signals	85
7	**Two-dimensional wave-packet interferometry for an electronic energy-transfer dimer**	**96**
	7.1 Energy-transfer dimer	96
	7.2 Whoopee signal	99

7.3 Illustrative calculations 105

Appendix A Electromagnetic energy change due to light absorption 129

Appendix B Delay regions for doubly excited-state-visiting overlaps in the difference-phased singly excited-state populations 131

Appendix C Delay regions for overlaps contributing to the difference-phased doubly excited-state population 135

Index 137

1
Short-pulse electronic absorption

1.1 Basic set-up

This chapter treats the interaction between an individual molecule, whose fixed location is regarded as the spatial origin, and a short pulse of light. The molecule may be isolated or immersed in a condensed-phase environment, but is assumed to have an energetically isolated electronic transition at or near resonance with the incident laser pulse. The time-dependent Hamiltonian of such a system can be written $H(t) = H + V(t)$, where

$$H = |g\rangle H_g \langle g| + |e\rangle(\epsilon + H_e)\langle e| \,, \tag{1.1}$$

and

$$V(t) = -\hat{\mathbf{m}} \cdot \mathbf{E}(t) \,. \tag{1.2}$$

H_g and H_e are the nuclear Hamiltonians governing all relevant intra- and intermolecular degrees of freedom in the ground and excited electronic states, respectively, and ϵ is the "bare" electronic transition energy. The electric dipole-moment operator[1] is

$$\hat{\mathbf{m}} = \mathbf{m}\big(|e\rangle\langle g| + |g\rangle\langle e|\big) \,, \tag{1.3}$$

and the electric field of the pulse is assumed to take the form

$$\mathbf{E}(t) = \mathbf{e} E f(t) \cos(\Omega t + \varphi) \,, \tag{1.4}$$

at the molecule's location, with polarization $\mathbf{e}$, pulse envelope $f(t)$ peaked at $t = 0$ with approximate temporal width σ, carrier frequency Ω, and unspecified optical phase φ.

The molecular state evolves according to

$$\frac{d}{dt}|\Psi(t)\rangle = \frac{1}{i\hbar} H(t)|\Psi(t)\rangle \,, \tag{1.5}$$

with the initial condition

$$|\Psi(t \ll 0)\rangle = [t]|g\rangle|n_g\rangle = |g\rangle[t]_{gg}|n_g\rangle = |g\rangle|n_g\rangle e^{-it\epsilon_{n_g}/\hbar} \,. \tag{1.6}$$

We have adopted the notation

[1]The electronic transition moment $\mathbf{m}$ may depend on nuclear coordinates in general, but we make a "Condon approximation" by neglecting this possibility for the present.

$$[t] \equiv \exp\{-iHt/\hbar\} \quad \text{and} \quad [t]_{gg} \equiv \langle g|\exp\{-iHt/\hbar\}|g\rangle = \exp\{-iH_g t/\hbar\}\,, \tag{1.7}$$

and $|n_g\rangle$ is a nuclear eigenket in the electronic ground state obeying $H_g|n_g\rangle = \epsilon_{n_g}|n_g\rangle$. We shall seek a perturbative solution through second order in V, which can be written abstractly as

$$|\Psi(t)\rangle = |0\rangle + |\uparrow\rangle + |\uparrow\downarrow\rangle\,. \tag{1.8}$$

Here $|0\rangle = [t]|g\rangle|n_g\rangle$ for all times; explicit formulas for $|\uparrow\rangle \cong |e\rangle\langle e|\Psi(t)\rangle$ and $|\uparrow\downarrow\rangle \cong |g\rangle\langle g|\Psi(t)\rangle - |0\rangle$ will be developed shortly.

1.2 Energy changes of molecule and field

In pulsed absorption, the molecular system experiences a change in energy

$$\begin{aligned}\Delta\mathcal{E} &= \langle\Psi(t \gg 0)|H|\Psi(t \gg 0)\rangle - \langle\Psi(t \ll 0)|H|\Psi(t \ll 0)\rangle \\ &\cong \langle\uparrow|H|\uparrow\rangle + \langle 0|H|\uparrow\downarrow\rangle + \langle\uparrow\downarrow|H|0\rangle \\ &= \langle\uparrow|H|\uparrow\rangle + 2\mathrm{Re}\langle 0|H|\uparrow\downarrow\rangle\,. \end{aligned} \tag{1.9}$$

By energy conservation, it must be that $\Delta\mathcal{E} + \Delta\mathcal{U} = 0$, where $\Delta\mathcal{U}$ is the accompanying energy change of the electromagnetic field, calculated to the same order in V. The electromagnetic energy change results from interference between the propagating incident field and the electric field $\boldsymbol{\mathcal{E}}(t)$ radiated by the oscillating dipole it induces in the molecule:

$$\begin{aligned}\Delta\mathcal{U} &= \frac{1}{4\pi}\int d^3s\big[(\mathbf{E}(t) + \boldsymbol{\mathcal{E}}(t))^2 - E^2(t)\big] \\ &\cong \frac{1}{2\pi}\int d^3s\,\mathbf{E}(t)\cdot\boldsymbol{\mathcal{E}}(t)\,; \end{aligned} \tag{1.10}$$

$\mathbf{E}(t)$ in this expression is obtained from that in eqn (1.4) by replacing t with $t - \mathbf{u}\cdot\mathbf{s}/c$, where $\mathbf{u}$ is a unit vector in the propagation direction and $\mathbf{s} = s\mathbf{n}$ locates the field-point. The radiated dipolar field is given by

$$\begin{aligned}\boldsymbol{\mathcal{E}}(t) &= \frac{1}{c^2 s}\big(\ddot{\mathbf{m}}(t - \tfrac{s}{c}) \times \mathbf{n}\big) \times \mathbf{n} \\ &= -\frac{1}{c^2 s}(\mathbf{1} - \mathbf{nn})\cdot\ddot{\mathbf{m}}(t - \tfrac{s}{c})\,; \end{aligned} \tag{1.11}$$

the double dots denote a second derivative with respect to time, and integration in eqn (1.10) is over the saucer-like spatial region where the two fields overlap, depicted in Fig. 1.1. The time-dependent dipole in eqn (1.11) is

$$\mathbf{m}(t) = \langle\Psi(t)|\hat{\mathbf{m}}|\Psi(t)\rangle = 2\mathrm{Re}\langle 0|\hat{\mathbf{m}}|\uparrow\rangle\,, \tag{1.12}$$

through first order in the external field. If we take z to be the direction of incidence of the laser pulse and x to be its polarization direction, then the energy change becomes

$$\Delta\mathcal{U} = -E\Omega\int_{-\infty}^{\infty} d\tau\, f(\tau)\sin(\Omega\tau + \varphi)m_x(\tau)\,; \tag{1.13}$$

this conversion from spatial to temporal integration is carried out in Appendix A.

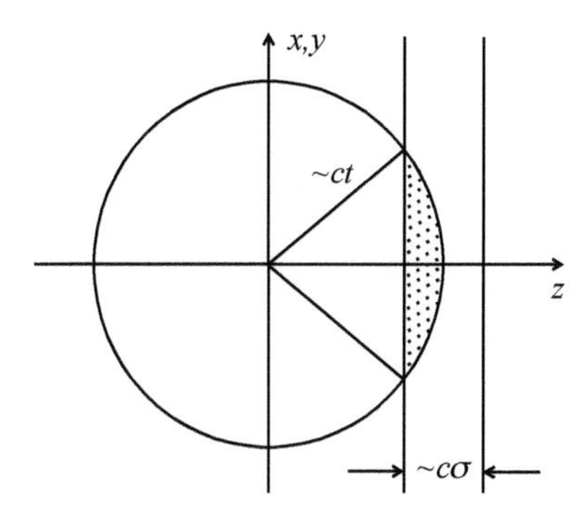

Fig. 1.1 Disk-shaped region of overlap between the propagating laser pulse and the induced dipolar field from a molecular source at the origin. Electromagnetic interference between these fields accounts for the light pulse's energy change due to its interaction with the molecule.

As hinted by the physically suggestive form of eqn (1.13), there is a direct route to this formula by way of a basic one for the rate of change of the molecular energy. In general, one has

$$\frac{d\mathcal{E}}{dt} = \frac{d}{dt}\langle\Psi(t)|H(t)|\Psi(t)\rangle = \langle\tfrac{d\Psi}{dt}|H(t)|\Psi\rangle + \langle\Psi|H(t)|\tfrac{d\Psi}{dt}\rangle + \langle\Psi|\tfrac{dH}{dt}|\Psi\rangle\,. \tag{1.14}$$

Using $|\frac{d\Psi}{dt}\rangle = \frac{1}{i\hbar}H(t)|\Psi\rangle$ and $H(t) = H + V(t)$, this becomes $d\mathcal{E}/dt = \langle\Psi|dV/dt|\Psi\rangle$; the well-known statement that the time derivative of the expectation value of the Hamiltonian equals the expectation value of the Hamiltonian's time derivative. In the present instance, we have

$$\frac{d\mathcal{E}}{dt} = -\frac{d\mathbf{E}(t)}{dt}\cdot\langle\Psi(t)|\hat{\mathbf{m}}|\Psi(t)\rangle \cong E\Omega f(t)\sin(\Omega t+\varphi)m_x(t)\,, \tag{1.15}$$

which, together with energy conservation, returns eqn (1.13) upon integration, irrespective of the order in E to which the induced dipole is approximated.

1.3 Expectation values

In order to relate $\Delta\mathcal{E}$ and $\Delta\mathcal{U}$ to the underlying quantum mechanical molecular dynamics, we need formulas for the zeroth-, first-, and second-order contributions to the time-dependent state (1.8). The corresponding interaction-picture ket, $|\tilde{\Psi}(t)\rangle \equiv [-t]|\Psi(t)\rangle$, obeys

$$\frac{d}{dt}|\tilde{\Psi}(t)\rangle = \frac{1}{i\hbar}[-t]\big(-H+H+V(t)\big)|\Psi(t)\rangle = \frac{1}{i\hbar}\tilde{V}(t)|\tilde{\Psi}(t)\rangle\,, \tag{1.16}$$

with $\tilde{V}(t) = [-t]V(t)[t]$, subject to the initial condition $|\tilde{\Psi}(t \ll 0)\rangle = |g\rangle|n_g\rangle$. The formal solution, $|\tilde{\Psi}(t)\rangle = |g\rangle|n_g\rangle + \frac{1}{i\hbar}\int_{-\infty}^{t} d\tau\,\tilde{V}(\tau)|\tilde{\Psi}(\tau)\rangle$, can be iterated to obtain a second-order approximation,

$$|\tilde{\Psi}(t)\rangle = \big\{1 + P(t;\tau) + P(t;\tau)P(\tau;\bar{\tau})\big\}|g\rangle|n_g\rangle\,, \tag{1.17}$$

where $P(t;\tau) \equiv \frac{1}{i\hbar}\int_{-\infty}^{t} d\tau\,\tilde{V}(\tau)$.[2]

For an electronically resonant or near-resonant field, it is a good approximation to neglect highly oscillatory terms in P's integrand (i.e., to make a rotating-wave approximation) and write

$$P(t;\tau) = iF\{|e\rangle\langle g|e^{-i\varphi}p^{(eg)}(t;\tau) + |g\rangle\langle e|e^{i\varphi}p^{(ge)}(t;\tau)\}\,, \tag{1.18}$$

where $F \equiv E\sigma\mathbf{e}\cdot\mathbf{m}/2\hbar$, and

$$p^{(eg)}(t;\tau) \equiv \int_{-\infty}^{t} \frac{d\tau}{\sigma}\,[-\tau]_{ee}[\tau]_{gg}f(\tau)e^{-i\Omega\tau} \tag{1.19}$$

is a *reduced pulse propagator*; here $[t]_{ee} = \langle e|[t]|e\rangle = \exp\{-i(\epsilon + H_e)t/\hbar\}$. It follows from this definition that $p^{(ge)} = (p^{(eg)})^{\dagger}$.

Using this notation in eqn (1.17) and reverting to the Schrödinger picture yield

$$\begin{aligned} |\Psi(t)\rangle = |g\rangle[t]_{gg}\{1 - F^2p^{(ge)}(t;\tau)p^{(eg)}(\tau;\bar{\tau})\}|n_g\rangle \\ + |e\rangle iFe^{-i\varphi}[t]_{ee}p^{(eg)}(t;\tau)|n_g\rangle\,, \end{aligned} \tag{1.20}$$

whence $|0\rangle = |g\rangle[t]_{gg}|n_g\rangle$ (as stated above),

$$|\uparrow\rangle = |e\rangle iFe^{-i\varphi}[t]_{ee}p^{(eg)}(t;\tau)|n_g\rangle\,, \tag{1.21}$$

and

$$|\uparrow\downarrow\rangle = -|g\rangle F^2[t]_{gg}p^{(ge)}(t;\tau)p^{(eg)}(\tau;\bar{\tau})|n_g\rangle\,. \tag{1.22}$$

The reduced pulse propagator $p^{(eg)}$ is evidently responsible for shaping the e-state nuclear wave packet generated from $|n_g\rangle$ in the $e \leftarrow g$ transition, while the nested combination $p^{(ge)}p^{(eg)}$ sculpts the second-order wave packet in the g-state.

Upon substituting these forms in eqn (1.9), we find

$$\begin{aligned} \Delta\mathcal{E} &= F^2\langle n_g|\,p^{(ge)}(\infty;\tau)(\epsilon + H_e)\,p^{(eg)}(\infty;\bar{\tau})|n_g\rangle \\ &\quad - 2F^2\,\mathrm{Re}\langle n_g|H_g\,p^{(ge)}(\infty;\tau)\,p^{(eg)}(\tau;\bar{\tau})|n_g\rangle \\ &= 2F^2\,\mathrm{Re}\langle n_g|\{\,p^{(ge)}(\infty;\tau)(\epsilon + H_e) - H_g\,p^{(ge)}(\infty;\tau)\}\,p^{(eg)}(\tau;\bar{\tau})|n_g\rangle\,; \end{aligned} \tag{1.23}$$

the observation time has been set to infinity because the energy increment stops changing after the end of the laser pulse. Notice that the molecular energy change is independent of the optical phase φ, as we should expect under the conditions considered here. The first term in the last member of eqn (1.23) accounts for the energy gained by population transfer to the electronic excited state, while the second tracks the energetic consequence of ground-state "bleaching" (i.e., the loss of g-state population).

[2]The τ appearing after the semicolon in $P(t;\tau)$ is not an argument of the pulse propagator, but simply identifies its integration variable. This notation proves useful in writing nested pulse propagators like those appearing in the second-order term of eqn (1.17).

For the change in electromagnetic energy (1.13) on the other hand, we have

$$\Delta\mathcal{U} = -E\Omega m_x \int_{-\infty}^{\infty} d\tau\, f(\tau)\sin(\Omega\tau+\varphi) \\ \times 2\mathrm{Re}\{iFe^{-i\varphi}\langle n_g|[-\tau]_{gg}[\tau]_{ee}p^{(eg)}(\tau;\bar{\tau})|n_g\rangle\}\,. \tag{1.24}$$

Within the rotating-wave approximation, this formula reduces to

$$\begin{aligned}\Delta\mathcal{U} &= -2F^2\hbar\Omega\,\mathrm{Re}\langle n_g|p^{(ge)}(\infty;\tau)p^{(eg)}(\tau;\bar{\tau})|n_g\rangle \\ &= -F^2\hbar\Omega\,\langle n_g|p^{(ge)}(\infty;\tau)p^{(eg)}(\infty;\bar{\tau})|n_g\rangle\,,\end{aligned} \tag{1.25}$$

which relates the field's energy loss to the center frequency of the laser pulse and the squared norm of the nuclear wave packet it generates in the excited electronic state.

Use $p^{(ge)}(\infty;\tau)(\epsilon+H_e) - H_g p^{(ge)}(\infty;\tau) = \int_{-\infty}^{\infty}\frac{d\tau}{\sigma}\big(i\hbar\frac{d}{d\tau}[-\tau]_{gg}[\tau]_{ee}\big)e^{i\Omega\tau}f(\tau)$ along with integration by parts to verify once again, acknowledging approximations, that $\Delta\mathcal{E} = -\Delta\mathcal{U}$, as required by energy conservation.

1.4 The Heller formula

We can explore the effect of pulse duration by adopting a specific, Gaussian form $f(\tau) = \exp\{-\tau^2/2\sigma^2\}$ for the envelope function. If we write

$$\Delta\mathcal{E} = \hbar\Omega F^2 \int_{-\infty}^{\infty}\frac{d\tau}{\sigma}\int_{-\infty}^{\infty}\frac{d\bar{\tau}}{\sigma}\, f(\tau)f(\bar{\tau})\langle n_g|\, e^{-\frac{i}{\hbar}(\epsilon+H_e-\epsilon_{n_g}-\hbar\Omega)(\tau-\bar{\tau})}|n_g\rangle\,, \tag{1.26}$$

and introduce new integration variables $T = (\tau+\bar{\tau})/2$ and $t = \tau-\bar{\tau}$, we then find

$$\Delta\mathcal{E} = \hbar\Omega F^2\sqrt{\pi}\int_{-\infty}^{\infty}\frac{dt}{\sigma}\, e^{i\Omega t}\langle n_g|\, e^{-\frac{i}{\hbar}(\epsilon+H_e-\epsilon_{n_g})t}|n_g\rangle\exp\{-t^2/4\sigma^2\}\,. \tag{1.27}$$

In the continuous-wave (cw) limit (σ longer than the inverse of the frequency spacing between adjacent vibronic levels), eqn (1.27) becomes Heller's celebrated expression for the molecular electronic absorption spectrum as a Fourier transform of the time-dependent overlap between a propagating e-state nuclear wave packet and the g-state wave function from which it originates.[3]

Equation (1.27) can be formally evaluated for shorter pulses (lower spectral resolution) by using e-state nuclear eigenkets obeying $H_e|n_e\rangle = \epsilon_{n_e}|n_e\rangle$ to obtain

$$\Delta\mathcal{E} = 2\pi\hbar\Omega F^2\sum_{n_e}|\langle n_e|n_g\rangle|^2\exp\{-\tfrac{\sigma^2}{\hbar^2}(\epsilon+\epsilon_{n_e}-\hbar\Omega-\epsilon_{n_g})^2\}\,. \tag{1.28}$$

As in the cw case, however, it is often more fruitful to evaluate $\Delta\mathcal{E}$ by treating the wave-packet overlap in Heller's formula under short-time, semiclassical, or system-bath-decomposition approximations. Reduced pulse propagators themselves can often

[3]Eric J. Heller, *The Semiclassical Way to Dynamics and Spectroscopy* (Princeton University Press, Princeton, 2018).

be handled with a variety of approximations relying on the brevity of ultrashort laser pulses relative to the timescale of nuclear dynamics.

1. Derive eqn (1.28) from eqn (1.27) by inserting the stated vibronic completeness relation and carrying out the resulting Gaussian integration.
2. Repeat this derivation, starting instead from the final form in eqn (1.25) and using $p^{(eg)}(\infty,\tau) = \sum_{n_e}\sum_{n_g} |n_e\rangle\langle n_g| \int_{-\infty}^{\infty} \frac{d\tau}{\sigma} f(\tau) e^{i\frac{\tau}{\hbar}(\epsilon+\epsilon_{n_e}-\hbar\Omega-\epsilon_{n_g})}$.

1.5 Systems starting in thermal equilibrium

In the treatment given so far, the initial state of the target system together with its environment is described "microcanonically," taking it to be of a specified-energy form proportional to $|g\rangle|n_g\rangle$. In this approach, the pulsed-absorption signal, along with the nonlinear optical signal expressions to be developed in subsequent chapters, is given as $\langle n_g|\hat{O}|n_g\rangle$, where $\hat{O}$ is a specified nuclear coordinate operator expressible in terms of wave-packet-reshaping reduced pulse propagators and episodes of evolution under the nuclear Hamiltonians associated with one or more electronic states; eqns (1.23) and (1.25) both have this structure.

It is to be appreciated that the signal formulas derived in this manner are much more general than the microcanonical set-up might seem to suggest. For the initial nuclear state $|n_g\rangle$ is assumed to be an eigenstate of H_g (not necessarily the ground state) and this Hamiltonian can be regarded as governing both "system" and "environmental" nuclear degrees of freedom: the former would comprise the translational, orientational, and internal vibrational modes of a target chromophore and may also include some nuclear degrees of freedom belonging to its immediate surroundings, while the latter would consist of the nuclear degrees of freedom of the remainder of the molecular environment.

If the system-environment boundary is defined spatially, for instance, and is set far enough away from the target chromophore that sound waves emanating from this molecule due to pulse-induced electronic transitions and the accompanying nuclear dynamics cannot propagate to the boundary and back on the overall timescale of the measurement (perhaps several tens of picoseconds), then the relevant portion of the nuclear operator $\hat{O} = \hat{O}_{sys} \otimes I_{env}$ pertains to the system alone. Reduced pulse propagators, in particular, take simplified forms such as $p^{(eg)} = p^{(eg)}_{sys} \otimes I_{env}$ to an excellent approximation, as the same condition automatically applies to the duration of any incident laser pulse. The signal can then be written as

$$\langle n_g|\hat{O}|n_g\rangle = \mathrm{Tr}_{sys}[\hat{O}_{sys}\hat{\rho}_{sys}], \tag{1.29}$$

where the quantum mechanical trace is taken over system degrees of freedom and $\hat{\rho}_{sys} = \mathrm{Tr}_{env}[\,|n_g\rangle\langle n_g|\,]$ is the system density operator expressed as a trace over environmental degrees of freedom of the microcanonical distribution.

If the "environment" is sufficiently large and sufficiently weakly coupled to the system that it becomes equivalent, as regards the statistical properties of the system, to a heat bath at some absolute temperature T, then the initial density operator of the system will take the canonical form $\hat{\rho}_{sys} = Z^{-1}\exp\{-H_g^{(sys)}/k_BT\}$. Here $Z =$

$\text{Tr}_{sys}[e^{-H_g^{(sys)}/k_BT}]$ is the partition function of the system and k_B is the Boltzmann constant. In this way, any of the optical signal expressions investigated here can be readily translated into the corresponding finite-temperature Boltzmann-weighted form.

A handy shortcut is to regard the $|n_g\rangle$ themselves as nuclear eigenstates of a subsystem in weak contact with a temperature reservoir. Then the thermally weighted sum of individual signal contributions $\langle n_g|\hat{O}|n_g\rangle$ reduces to $\text{Tr}[\hat{O}\hat{\rho}]$, with $\hat{\rho} = Z^{-1}\sum_{n_g}|n_g\rangle e^{-\epsilon_{n_g}/k_BT}\langle n_g|$, or $\text{Tr}[\hat{O}\hat{\rho}] = Z^{-1}\sum_{n_g} e^{-\epsilon_{n_g}/k_BT}\langle n_g|\hat{O}|n_g\rangle$.

1.6 Example absorption calculations

We can illustrate some basic features of pulsed linear absorption by investigating a model one-dimensional system with potential-energy curves reminiscent of a diatomic molecule embedded in a solid medium. The nuclear Hamiltonians in eqn (1.1) are taken to be $H_{g(e)} = \frac{p_x^2}{2m} + V_{g(e)}(x)$, where

$$V_g(x) = \frac{m\omega^2}{2}x^2 - \alpha x^3 + \beta x^4 \tag{1.30}$$

and $V_e(x) = V_g(x - x_e)$. Here $\alpha = \hbar\omega/32x_{rms}^3$ and $\beta = \hbar\omega/860x_{rms}^4$, with $x_{rms} = \sqrt{\hbar/2m\omega}$. We set the wave number of the harmonic vibration to $200\,\text{cm}^{-1}$, which corresponds to a period of $\tau_x = 2\pi/\omega = 166.8\,\text{fs}$. The spatial shift of the e-state potential is assigned the value $x_e = 2.626x_{rms}$, so the classical inner turning point at an energy ϵ_{2_e} lies at $x = 0$, directly above the g-state minimum. It is not necessary to specify a value for the vibrational reduced mass m. Fig. 1.2 shows the two potential-energy curves. The bare electronic transition energy ϵ in eqn (1.1) is given an arbitrary wave number of $20,000\,\text{cm}^{-1}$.

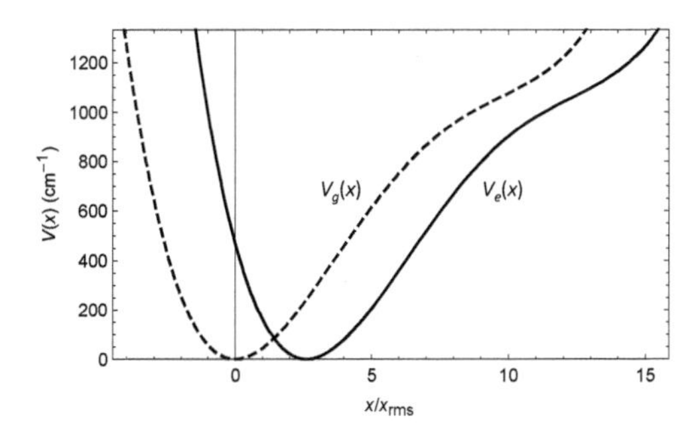

Fig. 1.2 Potential curves governing nuclear motion in the g- and e-states of the one-dimensional model Hamiltonian. V_e does not include the bare electronic transition energy.

For this simple system, it is a straightforward numerical exercise to solve for the eigenvalues and eigenstates of both nuclear Hamiltonians and to obtain position representations of the latter, using a harmonic-oscillator basis for instance. Matrix elements of the reduced pulse propagators can also be easily determined. The absorption spectrum can then be evaluated as a function of the carrier frequency for any incident pulse duration, using either eqn (1.25) or eqn (1.28). The spectrum for $\sigma = 5\tau_x$ starting from the initial state $|g\rangle|0_g\rangle$, shown in Fig. 1.3, is of sufficient resolution clearly to discern individual vibronic transitions. The absolute value of the corresponding Heller kernel appearing in the integrand of eqn (1.27) is plotted in the same figure, and reflects the complicated wave-packet dynamics that ensues upon the abrupt promotion of $|0_g\rangle$ to the e-state potential. It exhibits quasi-harmonic motion for a couple vibrational periods, after which wave-packet spreading in $V_e(x)$ born of anharmonicity gives rise to less regular temporal evolution in the overlap. The kernel's overall amplitude decays on a timescale set by the pulse-duration parameter σ.

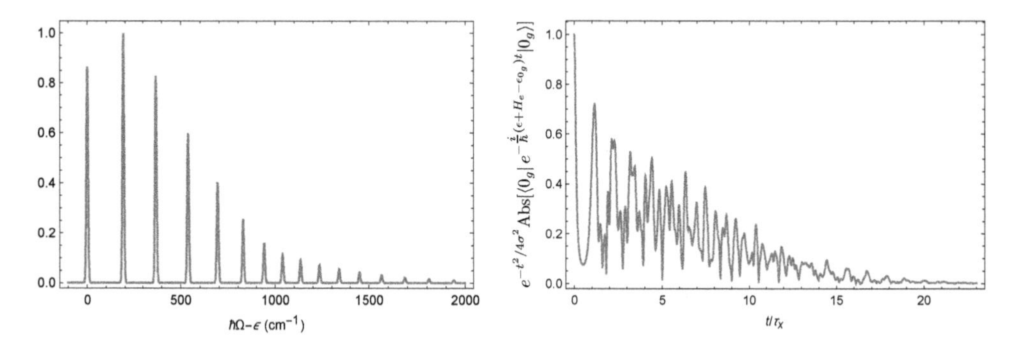

Fig. 1.3 On the left is the absorption spectrum for the 1D model, calculated for the case of narrow-bandwidth $\sigma = 5\tau_x$ pulses of continuously variable center frequency. The dimensionless field-strength parameter F is chosen so that the highest $(1_e \leftarrow 0_g)$ peak has unit intensity. The right panel shows the absolute value of the temporal overlap function, which provides a wave-packet-dynamical description of the absorption process.

The absorption spectrum of the same system at temperature $T = 298\,\text{K}$, with the same pulse duration and field strength, is plotted as a solid line in Fig. 1.4. While the spectral resolution is unchanged, the Boltzmann-weighted distribution of population over initial levels with $n_g = 0, 1, \ldots$ leads to the appearance of new peaks at various $n_e \leftarrow 1_g, 2_g, \ldots$ transition frequencies. The resolution *is* degraded with the use of wider-bandwidth, $\sigma = 0.5\tau_x$ pulses. The resulting linear absorption spectrum of the model system in the $n_g = 0$ state is shown as the dashed line in Fig. 1.4.

The spectra in Figs 1.3 and 1.4 illustrate several key ideas. The quasi-cw absorption of this relatively simple system provides detailed information on the energy eigenspectrum and the Franck–Condon overlaps $\langle n_e|n_g\rangle$, and is amenable to a dynamical description through the Heller kernel. In more complicated, multi-mode systems, however, thermal congestion and the presence of many closely spaced energy levels make it more difficult to resolve individual vibronic transitions. Short-pulse linear absorption, on the other hand, pays for its temporally well-localized excitation with a corresponding loss of spectral resolution. In a one-pulse experiment, this loss of resolution in the frequency domain remains uncompensated by improved temporal resolution resulting from precise control of the timing within a sequence of individual light-matter interactions. As is explored in subsequent chapters, strategies have been developed to address

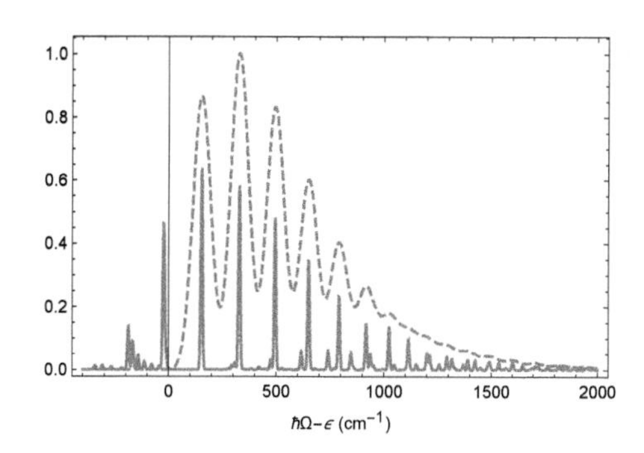

Fig. 1.4 Solid line is the absorption spectrum of the model at room temperature, again with $\sigma = 5.0\tau_x$. Dashed curve gives the spectrum starting from $n_g = 0$, with $\sigma = 0.5\tau_x$.

this conundrum by systematically varying the delay between excitation and probing in several different forms of ultrafast nonlinear optical spectroscopy.

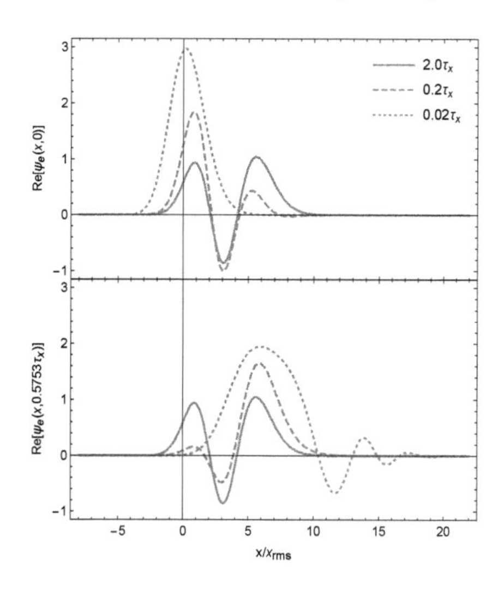

Fig. 1.5 Nuclear wave packets in the electronic excited state of the model system at $t = 0$ (top) and $0.5753\tau_x$ (bottom). The legend identifies the long, intermediate, and short values of the pulse-duration parameter σ used in these calculations

Figure 1.5 illustrates the dynamics initiated by pulses of differing duration in the model molecule. It shows wave packets in the position representation,

$$\psi_e(x,t) \equiv \langle x|e^{-i(H_e - \epsilon_{2_e})t/\hbar} p^{(eg)}(\infty;\tau)|0_g\rangle \,, \tag{1.31}$$

at $t = 0$ and $0.5753\tau_x$ (about half the period of vibrational motion at energy ϵ_{2_e}), with center frequency $\Omega = (\epsilon + \epsilon_{2_e} - \epsilon_{0_g})/\hbar$. The upper panel plots the real part of $\psi_e(x,0)$. For the longest pulse-length, $\sigma = 2.0\tau_x$, this initial wave packet closely resembles the stationary wave function $\langle x|2_e\rangle$. The shortest pulse, with $\sigma = 0.2\tau_x$, essentially copies $\langle x|0_d\rangle$ into the excited electronic state. Each of these wave packets after half a period of evolution, along with one generated by a pulse of intermediate duration, is illustrated in the lower panel. The longest-pulse packet remains almost unchanged, while that generated by the shortest pulse migrates to larger distance and is distorted by motion in the anharmonic e-state potential. It is ultrashort laser pulses' capacity to initiate and interrogate wave-packet dynamics that underlies their usefulness in time-resolved nonlinear optical spectroscopy.

2
Adiabatic approximation

2.1 Molecular Hamiltonian

The fact that atomic nuclei tend, as a result of their greater mass, to move more slowly than electrons do undergirds many basic features of molecular structure and dynamics. In particular, it motivates the ubiquitous Born–Oppenheimer, or adiabatic approximation, which greatly simplifies the quantum mechanical description of stationary and nonstationary molecular states. In this chapter, we describe the molecular adiabatic approximation and spell out the conditions for its validity.[1] In succeeding chapters, we will encounter additional applications of the quantum mechanical adiabatic theorem, which can be invoked whenever there is a clear timescale separation between relevant molecular degrees of freedom.

The time-independent Schrödinger for a molecule takes the general form

$$(T_N + T_e + V_{ee} + V_{eN} + V_{NN})|\Psi\rangle = E|\Psi\rangle\,, \tag{2.1}$$

in which the Hamiltonian operator is the sum of five terms:

$$T_N = \sum_{\text{nuclei } \alpha} \frac{\hat{\mathbf{P}}_\alpha \cdot \hat{\mathbf{P}}_\alpha}{2M_\alpha}$$

is the nuclear kinetic energy operator,

$$T_e = \sum_{\text{electrons } i} \frac{\hat{\mathbf{p}}_i \cdot \hat{\mathbf{p}}_i}{2m}$$

is the electronic kinetic energy,

$$V_{ee} = \sum_i \sum_{j>i} \frac{e^2}{|\hat{\mathbf{r}}_i - \hat{\mathbf{r}}_j|}$$

is the repulsive, electron-electron Coulombic potential energy,

$$V_{eN} = \sum_i \sum_\alpha \frac{(-Z_\alpha e)e}{|\hat{\mathbf{r}}_i - \hat{\mathbf{R}}_\alpha|}$$

[1]Some of the arguments in this section follow those made in Chap. 21 of Gordon Baym, *Lectures on Quantum Mechanics* (Westview Press, New York, 1990).

is the attractive, electron-nuclear Coulombic interaction, and

$$V_{NN} = \sum_{\alpha} \sum_{\beta > \alpha} \frac{(-Z_\alpha e)(-Z_\beta e)}{|\hat{\mathbf{R}}_\alpha - \hat{\mathbf{R}}_\beta|}$$

is the repulsive, nucleus-nucleus Coulombic interaction.

The ratio of electron mass to nuclear mass is typically $m/M \sim 10^{-4}$ to 10^{-5}. Typical electronic momenta are of size $h/\lambda_{deBroglie} \sim \hbar/a$, where a is the molecular size, so successive electronic energy levels are separated by about $\sim \hbar^2/ma^2$, or several electron volts. If a nucleus were displaced by the full molecular size a from its equilibrium position, the energy would change by an amount on the order of this electronic transition energy. Regarding the nuclear motion as roughly harmonic, then, $M\omega^2 a^2 \sim \hbar^2/ma^2$, which implies

$$\hbar\omega \sim \left(\frac{m}{M}\right)^{1/2} \frac{\hbar^2}{ma^2}\,. \tag{2.2}$$

Vibrational excitation energies are thereby found to be a smaller than electronic energies by a factor $(m/M)^{1/2} \sim 10^{-2}$. Additionally, since the nuclear kinetic energy is of size $P^2/2M \sim \hbar\omega/2$, nuclear momenta are roughly

$$P \sim \left(\frac{M}{m}\right)^{1/4} \frac{\hbar}{a} \sim \left(\frac{M}{m}\right)^{1/4} p\,. \tag{2.3}$$

Thus nuclear momenta are about ten times larger than electronic momenta, and nuclear *speeds* P/M are one thousand times smaller than electronic speeds. Finally, the typical deviation of a nuclear position from its equilibrium average value is given by $M\omega^2\delta^2/2 \sim \hbar\omega/2$, which leads to the estimate $(\delta/a)^2 \sim (m/M)^{1/2}$, or

$$\delta \sim \left(\frac{m}{M}\right)^{1/4} a\,; \tag{2.4}$$

the spatial range of zero-point nuclear motion is about one-tenth the range of electronic motion (i.e., one-tenth the "molecular size"). These ballpark figures support a physical picture of a molecule as a well-defined configuration of nuclei executing slow, small amplitude oscillations around their equilibrium positions within a delocalized medium of rapidly moving electrons.

Make a physical argument estimating molecular rotational excitation energies relative to vibrational and electronic energies.

2.2 Molecular eigenstates

Since the nuclei are well localized compared with the electrons—and much more slowly moving—while the nuclear excitation energies are a small fraction of electronic energies, we are led to regard the nuclear kinetic energy operator T_N as a perturbation and seek solutions to the *electronic Schrödinger equation* which obey

$$(T_e + V_{ee} + V_{eN} + V_{NN})|\varphi_n(\hat{R})\rangle = \mathcal{E}_n(\hat{R})|\varphi_n(\hat{R})\rangle\,. \tag{2.5}$$

The electronic eigenkets are parametrized by the nuclear coordinate *operators*, which we symbolically denote by $\hat{R} = (\hat{\mathbf{R}}_1, \hat{\mathbf{R}}_2, \dots)$. The electronic eigenenergies $\mathcal{E}_n(\hat{R})$ likewise depend on the locations of the nuclei.

We can relate the nuclear-coordinate-operator-dependent electronic eigenkets to those at some reference nuclear arrangement—say the equilibrium configuration in the electronic ground state—by a unitary tranformation:

$$|\varphi_n(\hat{R})\rangle = U(\hat{R})|\varphi_n\rangle\,. \tag{2.6}$$

The electronic unitary operator $U(\hat{R})$ is actually a large collection of different unitary operators parametrized by $\hat{R}$. Since the reference eigenkets $|\varphi_n\rangle$ constitute a complete orthonormal set in the electronic state-space, we could seek molecular eigenkets as expansions in this fixed electronic basis of the form $\sum_n |\varphi_n\rangle|\Theta_n\rangle$, where the $|\Theta_n\rangle$ are nuclear states. But it proves more useful by far to look for energy eigenstates of the combined electronic and nuclear degrees of freedom as expansions in the *adiabatic* electronic basis:

$$|\Psi\rangle = \sum_n |\varphi_n(\hat{R})\rangle|\Phi_n\rangle = U(\hat{R})\sum_n |\varphi_n\rangle|\Phi_n\rangle\,. \tag{2.7}$$

Justification for the adiabatic electronic basis as the practical choice emerges when we substitute $|\Psi\rangle$ in the molecular Schrödinger equation (2.1) and take account of the small ratio of electronic to nuclear mass.

Making this substitution and availing ourselves of the electronic Schrödinger equation (2.5) lead to

$$\sum_m \left[T_N + \mathcal{E}_m(\hat{R})\right]|\varphi_m(\hat{R})\rangle|\Phi_m\rangle = E\sum_m |\varphi_m(\hat{R})\rangle|\Phi_m\rangle\,. \tag{2.8}$$

Taking the inner product of eqn (2.8) with $|\varphi_n(\hat{R})\rangle$ yields a set of coupled equations for the nuclear states $\{|\Phi_n\rangle\}$:

$$\left[\langle\varphi_n(\hat{R})|T_N|\varphi_n(\hat{R})\rangle + \mathcal{E}_n(\hat{R})\right]|\Phi_n\rangle + \sum_{m\neq n}\langle\varphi_n(\hat{R})|T_N|\varphi_m(\hat{R})\rangle|\Phi_m\rangle = E|\Phi_n\rangle\,. \tag{2.9}$$

Let us examine the operator coupling nuclear kets associated with different electronic states, which can be rewritten as

$$\begin{aligned}\langle\varphi_n(\hat{R})|T_N|\varphi_m(\hat{R})\rangle &= \langle\varphi_n|U^\dagger(\hat{R})\sum_\alpha \frac{\hat{\mathbf{P}}_\alpha\cdot\hat{\mathbf{P}}_\alpha}{2M_\alpha}U(\hat{R})|\varphi_m\rangle \\ &= \langle\varphi_n|\sum_\alpha \frac{1}{2M_\alpha}U^\dagger(\hat{R})\hat{\mathbf{P}}_\alpha U(\hat{R})\cdot U^\dagger(\hat{R})\hat{\mathbf{P}}_\alpha U(\hat{R})|\varphi_m\rangle \\ &= \langle\varphi_n|\sum_\alpha \frac{1}{2M_\alpha}\left[\hat{\mathbf{P}}_\alpha + \mathbf{A}_\alpha(\hat{R})\right]\cdot\left[\hat{\mathbf{P}}_\alpha + \mathbf{A}_\alpha(\hat{R})\right]|\varphi_m\rangle\,, \end{aligned} \tag{2.10}$$

where the induced vector potential $\mathbf{A}_\alpha(\hat{R}) = -i\hbar U^\dagger(\hat{R})\nabla_\alpha U(\hat{R})$ is an operator in the electronic state-space parametrized by the nuclear coordinate operators.[2] Now

$$\begin{aligned}\langle\varphi_n(\hat{R})|T_N|\varphi_m(\hat{R})\rangle &= \langle\varphi_n|\varphi_m\rangle\sum_\alpha\frac{1}{2M_\alpha}\hat{\mathbf{P}}_\alpha\cdot\hat{\mathbf{P}}_\alpha \\ &+\sum_\alpha\frac{1}{2M_\alpha}\left(\hat{\mathbf{P}}_\alpha\cdot\langle\varphi_n|\mathbf{A}_\alpha(\hat{R})|\varphi_m\rangle+\langle\varphi_n|\mathbf{A}_\alpha(\hat{R})|\varphi_m\rangle\cdot\hat{\mathbf{P}}_\alpha\right) \\ &+\sum_\alpha\frac{1}{2M_\alpha}\langle\varphi_n|\mathbf{A}_\alpha(\hat{R})\cdot\mathbf{A}_\alpha(\hat{R})|\varphi_m\rangle\,. \end{aligned} \tag{2.11}$$

The inner product $\langle\varphi_n|\varphi_m\rangle$ equals δ_{nm}. The matrix elements of the vector potential, $\langle\varphi_n|\mathbf{A}_\alpha(\hat{R})|\varphi_m\rangle = -i\hbar\langle\varphi_n(\hat{R})|\nabla_\alpha\varphi_m(\hat{R})\rangle$, are of estimated size $\hbar/a$ as one would have to move the αth nucleus a molecule-sized distance to produce near unit overlap between a given adiabatic electronic state and another to which it had been orthogonal. The second sum on the right-hand side of eqn (2.11) therefore contains terms of order

$$\frac{\hbar}{M_\alpha a}P_\alpha \sim \frac{p^2}{m}\left(\frac{m}{M_\alpha}\right)^{3/4},$$

a thousand times smaller than electronic excitation energies. Terms in the third sum are of similar or smaller size,

$$\begin{aligned}\frac{1}{2M_\alpha}\langle\varphi_n|\mathbf{A}_\alpha(\hat{R})\cdot\mathbf{A}_\alpha(\hat{R})|\varphi_m\rangle &= \frac{1}{2M_\alpha}\sum_l\langle\varphi_n|\mathbf{A}_\alpha(\hat{R})|\varphi_l\rangle\cdot\langle\varphi_l|\mathbf{A}_\alpha(\hat{R})|\varphi_m\rangle \\ &\sim\frac{1}{2M_\alpha}\sum_l\left(\frac{\hbar}{a}\right)^2\sim 10\,\frac{m}{M_\alpha}\frac{p^2}{2m},\end{aligned}$$

assuming that ten states contribute significantly.

We see that the portions of T_N coupling different electronic states are about one one-thousandth of a standard electronic energy splitting. The corresponding perturbative corrections to a molecular eigenstate[3] go as

$$\frac{\text{matrix element}}{\text{electronic energy difference}}\sim 10^{-3},$$

while corrections to the eigenenergy are of size

$$\frac{(\text{matrix element})^2}{\text{energy difference}}\sim 10^{-6}\frac{\hbar^2}{ma^2}.$$

Except in the presence of degenerate or unusually closely spaced electronic energies, both of these corrections should be negligibly small.

[2]The gradient with respect to the αth nuclear coordinate is $\nabla_\alpha = \mathbf{i}\frac{\partial}{\partial X_\alpha}+\mathbf{j}\frac{\partial}{\partial Y_\alpha}+\mathbf{k}\frac{\partial}{\partial Z_\alpha}$. In evaluating $\nabla_\alpha U(\hat{R})$ it is to be understood that we first calculate $\nabla_\alpha U(\mathbf{R}_1,\mathbf{R}_2,\cdots)$ and then replace $(\mathbf{R}_1,\mathbf{R}_2,\cdots)$ with $(\hat{\mathbf{R}}_1,\hat{\mathbf{R}}_2,\cdots)$.

[3]See, for example, Chap. 5 of J. J. Sakurai, *Modern Quantum Mechanics*, Revised Edition (Addison-Wesley, Reading, 1994).

We are led to adopt the *Born–Oppenheimer approximation*, under which the expansion in eqn (2.7) is reduced to a single term,[4]

$$|\Psi\rangle \cong |\varphi_n(\hat{R})\rangle|\Phi_n\rangle\,. \tag{2.12}$$

In this approximation, the nuclear state that accompanies $|\varphi_n(\hat{R})\rangle$ obeys the equation

$$\left[\langle\varphi_n(\hat{R})|T_N|\varphi_n(\hat{R})\rangle + \mathcal{E}_n(\hat{R})\right]|\Phi_n\rangle = E|\Phi_n\rangle\,, \tag{2.13}$$

whose eigenvalues are the approximate molecular eigenenergies.

Using eqn (2.11) with m set equal to n and writing

$$\begin{aligned}\langle\varphi_n|\mathbf{A}_\alpha(\hat{R})\cdot\mathbf{A}_\alpha(\hat{R})|\varphi_n\rangle &= \langle\varphi_n|\mathbf{A}_\alpha(\hat{R})|\varphi_n\rangle\cdot\langle\varphi_n|\mathbf{A}_\alpha(\hat{R})|\varphi_n\rangle \\ &\quad + \sum_{m\neq n}\langle\varphi_n|\mathbf{A}_\alpha(\hat{R})|\varphi_m\rangle\cdot\langle\varphi_m|\mathbf{A}_\alpha(\hat{R})|\varphi_n\rangle\,,\end{aligned}$$

we can rewrite eqn (2.13) as

$$\left\{\sum_\alpha \frac{1}{2M_\alpha}\left[\hat{\mathbf{P}}_\alpha + \mathbf{A}_\alpha^{(n)}(\hat{R})\right]\cdot\left[\hat{\mathbf{P}}_\alpha + \mathbf{A}_\alpha^{(n)}(\hat{R})\right] + \mathcal{E}'_n(\hat{R})\right\}|\Phi_n\rangle = E|\Phi_n\rangle\,, \tag{2.14}$$

where

$$\mathbf{A}_\alpha^{(n)}(\hat{R}) = -i\hbar\langle\varphi_n|U^\dagger(\hat{R})\nabla_\alpha U(\hat{R})|\varphi_n\rangle = -i\hbar\langle\varphi_n(\hat{R})|\nabla_\alpha\varphi_n(\hat{R})\rangle \tag{2.15}$$

is the *adiabatic vector potential* experienced by the αth nucleus in the nth electronic state, and

$$\mathcal{E}'_n(\hat{R}) = \mathcal{E}_n(\hat{R}) + \sum_\alpha \frac{1}{2M_\alpha}\sum_{m\neq n}\langle\varphi_n|\mathbf{A}_\alpha(\hat{R})|\varphi_m\rangle\cdot\langle\varphi_m|\mathbf{A}_\alpha(\hat{R})|\varphi_n\rangle\,. \tag{2.16}$$

The nonadiabatic correction to $\mathcal{E}_n(\hat{R})$ is typically minuscule, as outlined above for $\langle\varphi_n|\mathbf{A}_\alpha(\hat{R})\cdot\mathbf{A}_\alpha(\hat{R})|\varphi_m\rangle/2M_\alpha$. The adiabatic vector potentials can often—but not always[5]—be made to vanish entirely: $\mathbf{A}_\alpha^{(n)}(\hat{R})$ must be *real*, for

$$\langle\varphi|\nabla\varphi\rangle = \nabla\underbrace{\langle\varphi|\varphi\rangle}_{1} - \langle\nabla\varphi|\varphi\rangle = -\langle\varphi|\nabla\varphi\rangle^*$$

is imaginary. If the electronic wave function $\langle\mathbf{r}_1,\mathbf{r}_2\cdots|\varphi_n(\hat{R})\rangle$ can be chosen real, as it most often can be, then $\langle\varphi_n(\hat{R})|\nabla_\alpha\varphi_n(\hat{R})\rangle$ will be real. Hence $\mathbf{A}_\alpha^{(n)}(\hat{R})$ is also *imaginary* and therefore zero.

[4]The Born–Oppenheimer-approximated state $|\Psi\rangle$ is not, in general, a tensor product. Rather it is a "sum" of products $\int d^3R_1 \int d^3R_2\cdots|\varphi_n(\mathbf{R}_1,\mathbf{R}_2,\cdots)\rangle|\mathbf{R}_1,\mathbf{R}_2,\cdots\rangle\langle\mathbf{R}_1,\mathbf{R}_2,\cdots|\Phi_n\rangle$, in which the electronic state is correlated with the arrangement of the nuclei and weighted by the probability amplitude, $\Phi_n(\mathbf{R}_1,\mathbf{R}_2,\cdots) = \langle\mathbf{R}_1,\mathbf{R}_2,\cdots|\Phi_n\rangle$, for the nuclei to have a particular arrangement.

[5]M. V. Berry, "Quantal phase factors accompanying adiabatic changes," Proc. R. Soc. Lond. **A392**, 45–57 (1984).

When the dust settles, it is often sufficient to combine a solution of the simplified nuclear wave equation,

$$\left[T_N + \mathcal{E}_n(\hat{R})\right] |\Phi_n\rangle = E|\Phi_n\rangle\,, \tag{2.17}$$

in which the electronic eigenenergy as a function of nuclear coordinates acts as a potential energy surface for the nuclear motion, with the adiabatic electronic state $|\varphi_n(\hat{R})\rangle$ to obtain an approximate molecular eigenket with total energy (electronic plus nuclear) equal to E.

1. How much can you figure out about the eigenstates and eigenenergies of the model Hamiltonian

$$H = \frac{P_1^2 + P_2^2}{2M} + \frac{M\omega^2}{2}(Q_1^2 + Q_2^2) + \kappa Q_1 \sigma_x + \kappa Q_2 \sigma_y\,,$$

where (Q_1, Q_2) and (P_1, P_2) are vibrational coordinates and momenta, respectively,

$$\sigma_x = |e\rangle\langle g| + |g\rangle\langle e|\,,$$

$$\sigma_y = -i|e\rangle\langle g| + i|g\rangle\langle e|\,,$$

and

$$\sigma_z = |e\rangle\langle e| - |g\rangle\langle e|\,,$$

are Pauli operators in a two-dimensional electronic state-space, and $\kappa = k\sqrt{\hbar M\omega^3}$, with a dimensionless constant k, is a coupling coefficient between electronic and nuclear degrees of freedom?

2. In addition to determining the stationary nuclear wave functions, the approximate Hamiltonian appearing in eqn (2.17) governs nuclear dynamics. Consider a one-dimensional vibrational Hamiltonian,

$$T + \mathcal{E}(X) = \frac{P^2}{2M} + \frac{M\omega^2}{2}X^2\,,$$

where $[X, P] = i\hbar$. Investigate the time-dependent expectation value and mean-square deviation of X and P in a *Glauber coherent state* specified by the initial condition $|\Phi\rangle = e^{-i\hat{P}X_\circ/\hbar}|0\rangle$, where $X_\circ$ is some initial displacement and $|0\rangle$ is the ground state of this harmonic nuclear Hamiltonian.

Two references pertinent to the first of these exercises are given below.[6]

[6]H. C. Longuet-Higgins, U. Öpik, M. H. L. Pryce, and R. A. Sack, "Studies of the Jahn-Teller effect. II. The dynamical problem," Proc. R. Soc. Lond. A **244**, 1–16 (1958); J. C. Slonczewski and V. L. Moruzzi, "Excited states in the dynamic Jahn-Teller effect," Physics **3**, 237–254 (1967).

3

Transient-absorption spectroscopy: Making ultrashort pulses worthwhile

This chapter develops a basic description of ultrafast transient-absorption spectroscopy and presents some calculated signals. By driving a sample with electronically resonant pump and delayed probe pulses, both of short duration on the timescale of atomic motion, and spectrally decomposing the transmitted probe beam, this widely applied technique overcomes short-pulse absorption's lack of spectral resolution in two respects: it gains high temporal resolution due to a precisely controlled time-delay between the pump and probe pulses, and it monitors spectral dynamics through frequency-by-frequency observation of the time-varying pump-induced intensity change of the transmitted probe. Transient-absorption spectroscopy provides direct information on wave-packet dynamics in more extended regions of the ground and excited electronic potential energy surfaces than linear absorption and, because it involves two pulses which may have different spectral composition, enables selective probing of multiple higher-lying states.

3.1 Model Hamiltonian and signal expression

We treat a molecule having three relevant electronic states, with the Hamiltonian

$$H = |g\rangle H_g \langle g| + |e\rangle (H_e + \epsilon_e) \langle e| + |f\rangle (H_f + \epsilon_f) \langle f| \,, \tag{3.1}$$

the potential energy surfaces of whose nuclear Hamiltonians take the value zero at their respective minima, and whose bare electronic energies obey $\epsilon_f - \epsilon_e < \epsilon_e < \epsilon_f$. The molecule's fixed location identifies the spatial origin. Its interaction with two ultrashort laser pulses is governed by

$$\begin{aligned} V(t) &= -\hat{m}E(t)\,; \\ \hat{m} &= m_{eg}(|e\rangle\langle g| + |g\rangle\langle e|) + m_{fe}(|f\rangle\langle e| + |e\rangle\langle f|)\,; \\ E(t) &= E_u f_u(t)\cos(\Lambda_u t + \varphi_u) + E_r f_r(t - t_d)\cos[\Lambda_r(t - t_d) + \varphi_r]\,. \end{aligned} \tag{3.2}$$

The pump (u) and probe (r) pulses have envelopes $f_u(t)$ and $f_r(t - t_d)$, respectively, of temporal width $\sim \sigma_u$ and $\sim \sigma_r$; their arrival times are separated by t_d. The two field strengths (E_u and E_r), center frequencies (Λ_u and Λ_r), and uncontrolled optical phases (φ_u and φ_r) are all independent.

The measured quantity in a transient-absorption experiment is the *change* in the spectrally resolved loss of energy from the probe pulse due to the prior action of the pump. To derive a formula for such a signal, we could calculate the change in electromagnetic energy—in a narrow frequency range of width $\delta\omega$ centered at $\bar{\omega}$—due to interference between the spatially propagating probe field and the field radiated by a portion of the system's induced dipole moment,

$$\begin{aligned}\Delta\mathcal{U}_{\bar{\omega}} &= \frac{1}{4\pi}\int d^3R\Big\{\big[\mathbf{E}_{r\bar{\omega}}(\mathbf{R},t)+\mathcal{E}_{u^2r\bar{\omega}}(\mathbf{R},t)\big]\cdot\big[\mathbf{E}_{r\bar{\omega}}(\mathbf{R},t)+\mathcal{E}_{u^2r\bar{\omega}}(\mathbf{R},t)\big] \\ &\quad - E^2_{r\bar{\omega}}(\mathbf{R},t)\Big\} \\ &\cong \frac{1}{2\pi}\int d^3R\,\mathbf{E}_{r\bar{\omega}}(\mathbf{R},t)\cdot\mathcal{E}_{u^2r\bar{\omega}}(\mathbf{R},t)\,. \end{aligned} \tag{3.3}$$

In this equation, $\mathcal{E}_{u^2r}$ is the contribution to the induced dipolar field proportional to pump *intensity* and the probe *field strength.* The subscript $\bar{\omega}$ denotes a spectrally filtered field component; for instance, the filtered probe is given by

$$\mathbf{E}_{r\bar{\omega}}(\mathbf{R},t) = \int_{\bar{\omega}-\frac{\delta\omega}{2}}^{\bar{\omega}+\frac{\delta\omega}{2}} \frac{d\omega}{2\pi}\,e^{-i\omega t}\tilde{\mathbf{E}}_r(\mathbf{R},\omega) + \text{c.c.}\,, \tag{3.4}$$

where $\tilde{\mathbf{E}}_r(\mathbf{R},\omega) = \int_{-\infty}^{\infty} dt\, e^{i\omega t}\mathbf{E}_r(\mathbf{R},t)$.[1]

The integration over space in eqn (3.3) could be evaluated by a procedure analogous to that in Appendix A.[2] As an alternative, we pursue a direct derivation of the transient-absorption signal as a time integral for the contribution to the change in molecular energy, $\Delta\mathcal{E}_{\bar{\omega}} = -\Delta\mathcal{U}_{\bar{\omega}}$, that is bilinear in the pulse intensities. As in eqn (3.2), we assume for simplicity that both transition dipole moments and both field polarizations are oriented along the same (say x-) axis. Temporarily writing $m(t)$ for $m_{u^2r}(t)$ and $E(t)$ for $E_r(t)$, we can break up the probe field into pieces which would be detected by idealized pixels sensitive to different small frequency ranges:

$$E(t) = \sum_{\bar{\omega}>0} E_{\bar{\omega}}(t)\,, \tag{3.5}$$

and the transient-absorption dipole could be similarly decomposed as

$$m(t) = \sum_{\bar{\omega}>0} m_{\bar{\omega}}(t)\,. \tag{3.6}$$

In terms of the latter two quantities, the transient-absorption signal would be written

$$\Delta\mathcal{E}_{\bar{\omega}} = -\int_{-\infty}^{\infty} dt\,\frac{dE_{\bar{\omega}}(t)}{dt}m_{\bar{\omega}}(t)\,, \tag{3.7}$$

in which both probe-field and the dipole-moment components belong to the spectral segment centered at $\bar{\omega}$ (compare eqn (1.15)). Now,

[1]Equations (3.3) and (3.4) differ from eqn (3.2) in making explicit the vector nature of the incident and radiated fields.

[2]See also J. A. Cina, P. A. Kovac, C. C. Jumper, J. C. Dean, and G. D. Scholes, "Ultrafast transient absorption revisited: Phase-flips, spectral fingers, and other dynamical features," J. Chem. Phys. **144**, 175102/1–18 (2016).

$$E_{\bar{\omega}}(t) = \int_{\bar{\omega}-\frac{\delta\omega}{2}}^{\bar{\omega}+\frac{\delta\omega}{2}} \frac{d\omega}{2\pi}\, e^{-i\omega t}\tilde{E}(\omega) + \text{c.c.}\,, \tag{3.8}$$

where

$$\tilde{E}(\omega) = \int_{-\infty}^{\infty} dt\, e^{i\omega t} E(t) = e^{i\omega t_d} \int_{-\infty}^{\infty} d\tau\, e^{i\omega\tau} E(\tau + t_d)\,. \tag{3.9}$$

Since the probe field $E(\tau + t_d)$ has a temporal width $\sim\sigma \ll 2\pi/\delta\omega$ about $\tau = 0$, for frequencies in the range $\bar{\omega} - \frac{\delta\omega}{2} < \omega < \bar{\omega} + \frac{\delta\omega}{2}$ we have

$$\tilde{E}(\omega) \cong e^{i\omega t_d} \int_{-\infty}^{\infty} d\tau\, e^{i\bar{\omega}\tau} E(\tau + t_d) = e^{i(\omega-\bar{\omega})t_d}\tilde{E}(\bar{\omega})\,. \tag{3.10}$$

As a result, from eqn (3.8),

$$E_{\bar{\omega}}(t) \cong 2\,\text{Re}\big\{\tilde{E}(\bar{\omega})e^{-i\bar{\omega}t}\big\}\, \frac{\sin\frac{\delta\omega}{2}(t - t_d)}{\pi(t - t_d)}\,. \tag{3.11}$$

The molecular energy acquisition involving a particular spectral segment of the probe field and the *full* induced dipole moment is

$$-\int_{-\infty}^{\infty} dt\, \frac{dE_{\bar{\omega}}(t)}{dt} m(t) \cong i\bar{\omega}\frac{\delta\omega}{2\pi}\tilde{E}(\bar{\omega})\tilde{m}^*(\bar{\omega}) + \text{c.c.} \tag{3.12}$$

Since the right-hand side of eqn (3.12) involves the $\bar{\omega}$ Fourier components of *both* the probe field and the induced dipole, it clearly corresponds to $\Delta\mathcal{E}_{\bar{\omega}}$ of eqn (3.7), the transient-absorption signal at the given frequency. Reintroducing the appropriate subscripts, the signal may therefore be written

$$\Delta\mathcal{E}_{\bar{\omega}} = -\int_{-\infty}^{\infty} dt\, \frac{dE_{r\bar{\omega}}(t)}{dt} m_{u^2r}(t)\,; \tag{3.13}$$

the transient-absorption dipole in its entirety appears in the integrand, rather than just its $\bar{\omega}$ element.

For a probe pulse of the form given in eqn (3.2), the Fourier components are

$$\begin{aligned} \tilde{E}_r(\omega > 0) &\cong \frac{E_r}{2} e^{i\omega t_d - i\varphi_r} \int_{-\infty}^{\infty} dt\, e^{i(\omega-\Lambda_r)(t-t_d)} f_r(t - t_d) \\ &= \frac{E_r}{2} e^{i\omega t_d - i\varphi_r} \tilde{f}_r(\omega - \Lambda_r)\,, \end{aligned} \tag{3.14}$$

whence[3]

$$E_{r\bar{\omega}}(t) \cong E_r \tilde{f}_r(\bar{\omega} - \Lambda_r) \frac{\sin\frac{\delta\omega}{2}(t - t_d)}{\pi(t - t_d)} \cos[\bar{\omega}(t - t_d) + \varphi_r]\,. \tag{3.15}$$

Inserting eqn (3.15) in eqn (3.13) gives a working formula for the transient-absorption signal,

[3]We specialize to symmetric pulse envelopes, so $\tilde{f}_r(\bar{\omega} - \Lambda_r)$ is real-valued.

$$\Delta\mathcal{E}_{\bar{\omega}} = \bar{\omega}E_r\tilde{f}_r(\bar{\omega}-\Lambda_r)\int_{-\infty}^{\infty} dt\, m_{u^2r}(t)\frac{\sin\frac{\delta\omega}{2}(t-t_d)}{\pi(t-t_d)}\sin[\bar{\omega}(t-t_d)+\varphi_r]\,. \tag{3.16}$$

The dynamics of the targeted system enters the signal through the interpulse delay-dependence and spectral content of the induced dipole moment.

Fill in the steps that lead to eqns (3.11), (3.12), and (3.15). Show that eqn (3.11) is consistent with eqn (3.15) for a probe pulse of the form given in eqn (3.2).

3.2 Transient-absorption dipole

3.2.1 Contributions to m_{u^2r}

In order to calculate m_{u^2r}, the portion of $\langle\Psi(t)|\hat{m}|\Psi(t)\rangle$ that is of second order in the pump field and first order in the probe, we need a perturbative expansion of the molecular state including all relevant terms through that same order. In a system with three electronic states of the variety under study, this expansion can be written

$$\begin{aligned}|\Psi\rangle \cong |0\rangle &+ |\uparrow_u\rangle + |\uparrow_r\rangle + |\uparrow_u\uparrow_u\rangle + |\uparrow_u\downarrow_u\rangle + |\uparrow_u\uparrow_r\rangle + |\uparrow_u\downarrow_r\rangle + |\uparrow_r\uparrow_u\rangle + |\uparrow_r\downarrow_u\rangle \\ &+ |\uparrow_r\uparrow_r\rangle + |\uparrow_r\downarrow_r\rangle + |\uparrow_u\uparrow_u\downarrow_r\rangle + |\uparrow_u\downarrow_u\uparrow_r\rangle + |\uparrow_u\uparrow_r\downarrow_u\rangle + |\uparrow_u\downarrow_r\uparrow_u\rangle \\ &+ |\uparrow_r\uparrow_u\downarrow_u\rangle + |\uparrow_r\downarrow_u\uparrow_u\rangle\,. \end{aligned} \tag{3.17}$$

Here $|0\rangle$ is the time-dependent molecular state unperturbed by either laser pulse. The arrows in the other kets signify energetically upward (g-to-e or e-to-f) or downward (e-to-g or f-to-e) electronic transitions driven by the pump or the probe, as indicated by the subscript. These transitions are required to occur in the order listed from left to right. Kets in which the pulses act "out of order" (probe before pump) vanish for interpulse delays significantly longer than the pulse durations. From the expectation value of the dipole-moment operator, we can isolate

$$\begin{aligned} m_{u^2r}(t) = &\underbrace{\langle 0|\hat{m}|\uparrow_u\uparrow_u\downarrow_r\rangle}_{\exp\{-2i\varphi_u\}} + \langle\uparrow_u\uparrow_r\downarrow_u|\hat{m}|0\rangle + \langle\uparrow_r\uparrow_u\downarrow_u|\hat{m}|0\rangle \\ &+ \underbrace{\langle\uparrow_u\uparrow_r|\hat{m}|\uparrow_u\rangle}_{\text{ESA}} + \langle\uparrow_r\uparrow_u|\hat{m}|\uparrow_u\rangle + \underbrace{\langle\uparrow_r|\hat{m}|\uparrow_u\uparrow_u\rangle}_{\exp\{-2i\varphi_u\}} \\ &+ \underbrace{\langle\uparrow_u\downarrow_u\uparrow_r|\hat{m}|0\rangle}_{\text{GSB}} + \underbrace{\langle 0|\hat{m}|\uparrow_u\downarrow_r\uparrow_u\rangle}_{\exp\{-2i\varphi_u\}} + \langle\uparrow_r\downarrow_u\uparrow_u|\hat{m}|0\rangle \\ &+ \underbrace{\langle\uparrow_r|\hat{m}|\uparrow_u\downarrow_u\rangle}_{\text{ISRS}} + \underbrace{\langle\uparrow_u|\hat{m}|\uparrow_u\downarrow_r\rangle}_{\text{SE}} + \underbrace{\langle\uparrow_r\downarrow_u|\hat{m}|\uparrow_u\rangle}_{\exp\{-2i\varphi_u\}} + \text{c.c.} \end{aligned} \tag{3.18}$$

The four under-braced matrix elements will be seen to carry an uncontrolled optical phase factor $\exp\{-2i\varphi_u\}$. Over the many laser shots needed to acquire a transient-absorption signal, the phase $2\varphi_u$ randomly samples the range from zero to 2π, leading to a negligible net contribution from these elements.

In the terms in eqn (3.18) labeled excited-state absorption (ESA), ground-state bleach (GSB), impulsive stimulated Raman scattering (ISRS), and stimulated emission (SE), none of the pulse actions listed separately within the bra or the ket occur outside their nominal order (pump before probe). Only these terms survive when t_d significantly exceeds the pulse durations. The four remaining elements can contribute only at shorter delays. Two of these correspond to variations of the ESA and GSB elements in which the probe interchanges roles with one of the pump actions; in the absence of different spectral content in the pump and probe, the molecule itself would not distinguish these terms from their conventionally interpreted counterparts. The same relationship exists between the SE and ISRS terms. The two other short-t_d elements are optical-phase-controlled counterparts of the stricken, unstable element $\langle 0|\hat{m}|\uparrow_u\uparrow_u\downarrow_r\rangle$.

Provide physical explanations for each of the interpretive labels in eqn (3.18).

3.2.2 Multi-pulse bras and kets

Time-dependent perturbation theory provides formulas for the various multi-pulse contributions to the molecular state in eqn (3.17) which enter the induced dipole of eqn (3.18). We seek a solution to $i\hbar\frac{\partial}{\partial t}|\Psi(t)\rangle = \big(H + V(t)\big)|\Psi(t)\rangle$ with the initial condition $|\Psi(t \ll 0)\rangle = [t]|g\rangle|n_g\rangle$, where $H_g|n_g\rangle = \epsilon_{n_g}|n_g\rangle$ (consult Chapter 1 for notation). Through third order in V, we find

$$|\Psi(t)\rangle = \Bigg\{[t] + i\sum_{j=u,r}[t-t_j]P_j(t-t_j;\tau)[t_j] \tag{3.19}$$
$$+ i^2\sum_{j,k}[t-t_j]P_j(t-t_j;\tau)[t_j-t_k]P_k(\tau+t_j-t_k;\bar{\tau})[t_k]$$
$$+ i^3\sum_{j,k,l}[t-t_j]P_j(t-t_j;\tau)[t_j-t_k]P_k(\tau+t_j-t_k;\bar{\tau})[t_k-t_l]P_l(\bar{\tau}+t_k-t_l;\bar{\bar{\tau}})[t_l]\Bigg\}|g\rangle|n_g\rangle\,,$$

where $t_u = 0$, $t_r = t_d$, and

$$P_j(t;\tau) = F_j^{(eg)}e^{-i\varphi_j}|e\rangle\langle g|\,p_j^{(eg)}(t;\tau) + F_j^{(fe)}e^{-i\varphi_j}|f\rangle\langle e|\,p_j^{(fe)}(t;\tau) + \text{H.c.}\,, \tag{3.20}$$

after making a rotating-wave approximation. In eqn (3.20), we have used reduced pulse propagators of the form

$$p_j^{(eg)}(t;\tau) = \int_{-\infty}^{t}\frac{d\tau}{\sigma_j}\,[-\tau]_{ee}[\tau]_{gg}e^{-i\Lambda_j\tau}f_j(\tau)\,, \tag{3.21}$$

for example, along with $F_j^{(eg)} = E_j m_{eg}\sigma_j/2\hbar$.

Rather than generate explicit expressions for all the multi-pulse kets and bras appearing in eqn (3.18), some of which may become insignificant with particular choices of pulse center frequencies and bandwidths, or certain ranges of interpulse delay time, it is more instructive simply to write out a few examples. We have, for instance,

$$|0\rangle = [t]|g\rangle|n_g\rangle = |g\rangle[t]_{gg}|n_g\rangle = |g\rangle|n_g\rangle e^{-i\epsilon_{n_g}t/\hbar}\,; \tag{3.22}$$

$$|\uparrow_u\rangle = |e\rangle iF_u^{(eg)}e^{-i\varphi_u}[t]_{ee}\,p_u^{(eg)}(t;\tau)|n_g\rangle\,; \tag{3.23}$$

$$|\uparrow_r\rangle = |e\rangle iF_r^{(eg)}e^{-i\varphi_r}[t-t_d]_{ee}\,p_r^{(eg)}(t-t_d;\tau)[t_d]_{gg}|n_g\rangle\,; \tag{3.24}$$

and

$$|\uparrow_u\uparrow_r\rangle = |f\rangle i^2F_u^{(eg)}F_r^{(fe)}e^{-i\varphi_u-i\varphi_r}[t\text{-}t_d]_{ff}\,p_r^{(fe)}(t\text{-}t_d;\tau)[t_d]_{ee}\,p_u^{(eg)}(\tau+t_d;\bar{\tau})|n_g\rangle\,. \tag{3.25}$$

Gain some practice by writing expressions analogous to eqns (3.22)–(3.25) for a few more multi-pulse kets.

From expressions of the sort just considered, we can obtain the dipole matrix elements. For the ESA element, for example, we find

$$\begin{aligned}\langle\uparrow_u\uparrow_r|\hat{m}|\uparrow_u\rangle = -im_{fe}\big(F_u^{(eg)}\big)^2F_r^{(fe)}e^{i\varphi_r}\langle n_g|p_u^{(ge)}(\tau+t_d;\bar{\tau})[-t_d]_{ee}\,p_r^{(ef)}(t-t_d;\tau)\\ \times[-t+t_d]_{ff}[t-t_d]_{ee}\,[t_d]_{ee}\,p_u^{(eg)}\big((t-t_d)+t_d;\bar{\bar{\tau}}\big)|n_g\rangle\,.\end{aligned} \tag{3.26}$$

If we define a filtered-probe reduced pulse propagator,

$$p_{r\bar{\omega}}^{(fe)}(\infty;\tau) = \int_{-\infty}^{\infty}\frac{d\tau}{\bar{\sigma}}\,\frac{\sin\frac{\delta\omega}{2}\tau}{\frac{\delta\omega}{2}\tau}[-\tau]_{ff}[\tau]_{ee}\,e^{-i\bar{\omega}\tau}\,, \tag{3.27}$$

with $\bar{\sigma}\equiv 2\pi/\delta\omega$, write $\sin[\bar{\omega}(t-t_d)+\varphi_r] = \frac{i}{2}\exp\{-\,i\bar{\omega}(t-t_d)-i\varphi_r\}+\text{c.c.}$ in eqn (3.16), and neglect terms in its integrand that oscillate at optical frequencies, we then find the contribution of excited-state absorption to the transient-absorption signal,

$$\begin{aligned}\Delta\mathcal{E}_{\bar{\omega}\mathrm{ESA}} = \hbar\bar{\omega}\frac{\tilde{f}_r(\bar{\omega}-\Lambda_r)}{\sigma_r}\big(F_u^{(eg)}\big)^2\big(F_r^{(fe)}\big)^2\langle n_g|p_u^{(ge)}(\tau+t_d;\bar{\tau})[-t_d]_{ee}\,p_r^{(ef)}(t;\tau)\\ \times\,p_{r\bar{\omega}}^{(fe)}(\infty;t)[t_d]_{ee}\,p_u^{(eg)}(t+t_d;\bar{\bar{\tau}})|n_g\rangle+\text{c.c.}\,,\end{aligned} \tag{3.28}$$

where $\tilde{f}_r(\omega) = \int_{-\infty}^{\infty}dt\,e^{i\omega t}f_r(t)$, as defined in eqn (3.14). Notice is to be taken of the t-nesting that occurs because the reduced propagator for the spectrally filtered probe pulse is the last-acting operator. Similar expressions for the other transient-absorption signal contributions can readily be found by analogy with eqn (3.28).

Carry out this exercise for the contributions of ground-state bleach, impulsive stimulated Raman scattering, and stimulated emission to transient absorption. You can check your answers by consulting Section 3.3.2.

3.3 Exemplary calculations

3.3.1 2D model and absorption spectrum

In order to carry out illustrative calculations of transient-absorption signals, we frame a model Hamiltonian in the form of eqn (3.1) for a molecular system consisting of one slower and one faster nuclear degree of freedom, which could be thought of, respectively, as a slow "conformational" coordinate and an internal molecular vibration whose equilibrium average value is slightly displaced upon electronic excitation. The three nuclear Hamiltonians are

$$H_j = \frac{P^2}{2} + \frac{p^2}{2} + V_j(X, x)\,, \tag{3.29}$$

for $j = g, e, f$. The g-state potential is

$$V_g(X, x) = \frac{\Omega^2}{2}(X - X_g)^2 + \frac{\omega_g^2}{2}(x - x_g)^2\,. \tag{3.30}$$

The slow-mode frequency Ω serves as a reference, set to 40 cm^{-1} for plotting purposes, while $\omega_g = 13.5\,\Omega$; the ground-state minimum lies at $(X_g, x_g) = (-1.8\,X_{rms}, 0\,x_{rms})$, where $X_{rms} = \sqrt{\hbar/2\Omega}$ and $x_{rms} = \sqrt{\hbar/2\omega_g}$. The slow mode serves as a photochemical reaction coordinate in the e-state, whose potential function is

$$\begin{aligned} V_e(X, x) &= \alpha_e\left(\frac{3}{2}(X - X_e)^4 + 4X_e(X - X_e)^3 + 3X_e^2(X - X_e)^2\right) \\ &\quad + \frac{\omega_e^2}{2}(x - x_e)^2 + \beta_e(X - X_e)(x - x_e)\,, \end{aligned} \tag{3.31}$$

with $\alpha_e = 0.01613\,\hbar\Omega/X_{rms}^4$, $X_e = 4.0\,X_{rms}$, $\omega_e = 11.0\,\Omega$, $x_e = 1.5\,x_{rms}$, and $\beta_e = -0.4550\,\hbar\Omega/X_{rms}x_{rms}$. We set the bare electronic energy of this state to $\epsilon_e = 600.0\,\hbar\Omega$. The f-state potential function is of similar form,

$$\begin{aligned} V_f(X, x) &= \alpha_g\left(\frac{3}{2}(X - X_f)^4 + 4X_f(X - X_f)^3 + 3X_f^2(X - X_f)^2\right) \\ &\quad + \frac{\omega_f^2}{2}(x - x_f)^2 + \beta_f(X - X_f)(x - x_f)\,, \end{aligned} \tag{3.32}$$

in which $\alpha_f = 0.02419\,\hbar\Omega/X_{rms}^4$, $X_f = 3.0\,X_{rms}$, $\omega_f = 9.0\,\Omega$, $x_f = 0.9\,x_{rms}$, and $\beta_f = -0.3344\,\hbar\Omega/X_{rms}x_{rms}$. The bare electronic energy of the f-state is assigned the value $\epsilon_f = 900.0\,\hbar\Omega$. Contour diagrams of the three electronic potential energy surfaces are displayed in Fig. 3.1.

The continuous-wave electronic absorption spectrum provides a first acquaintance with the spectroscopy and dynamics of this system. The model Hamiltonian is simple enough to be diagonalizable (we used a discrete position basis), and the spectrum can readily be calculated from eqn (1.28) once the transition energies and Franck–Condon overlaps are determined. As seen in Fig. 3.2, with the incident frequency denoted as Λ and line-widths determined by the finite duration $\sigma = 1.6 \times 2\pi/\Omega$ of a Gaussian pulse,

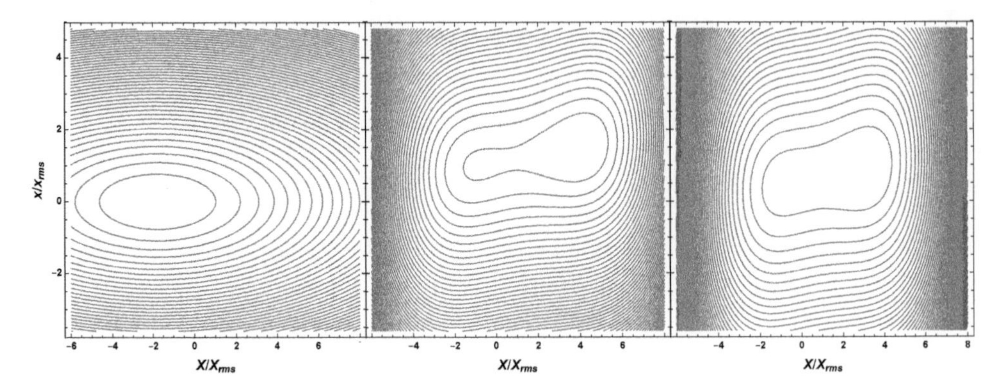

Fig. 3.1 V_g, V_e, and V_f are shown from left to right. Contours are separated by $2\,\hbar\Omega$.

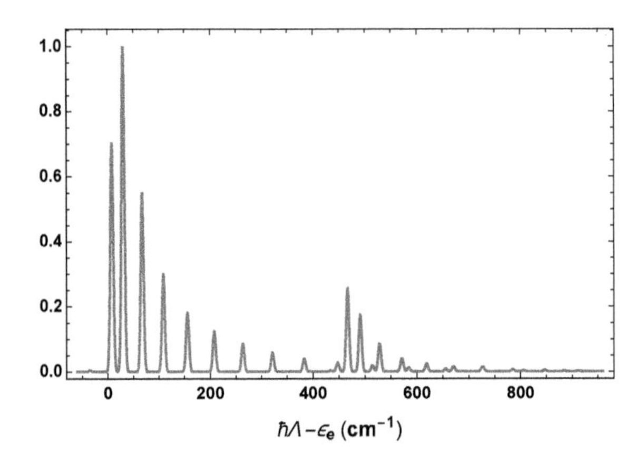

Fig. 3.2 Linear absorption of the two-dimensional model system in arbitrary units versus wave-number displacement from the $e \leftarrow g$ bare electronic transition energy.

the spectrum exhibits interesting progressions in both the low-frequency, reaction-coordinate mode and the higher-frequency vibration. The transition at $\epsilon_e - 34.4\,\text{cm}^{-1}$ from the initially populated lowest level in the g-state to the photochemical "product state" corresponding to the lowest e-state level—localized in the potential-energy well at (X_e, x_e)—is barely visible due to the tiny Franck–Condon overlap between the associated nuclear wave functions.

Can you rationalize the value $\epsilon_e - 34.4\,\text{cm}^{-1}$ of the energy for the very weak zero-zero absorption transition through a consideration of the zero-point energies in the ground and excited electronic states?

The wave-packet dynamics underlying linear absorption in this system is brought to light by the Heller overlap,

$$C(t) = \langle 0_g | e^{itH_g/\hbar} e^{-it(\epsilon_e + H_e)/\hbar} | 0_g \rangle, \tag{3.33}$$

appearing in eqn (1.27). In this formula, $|0_g\rangle$ denotes the two-dimensional vibrational ground state. As explained in Chapter 1, $C(t)$ is the complex-valued overlap between the initial, stationary vibrational wave function and the moving wave packet engendered by its transfer to the electronic excited state. Fourier transformation of eqn (3.33) times the temporal window $\exp\{-t^2/4\sigma^2\}$ (and some constant pre-factors) yields the spectrum. The absolute value of the overlap times the window function is plotted in Fig. 3.3. It decays rapidly as the 2D wave packet gains slow-mode momentum P and

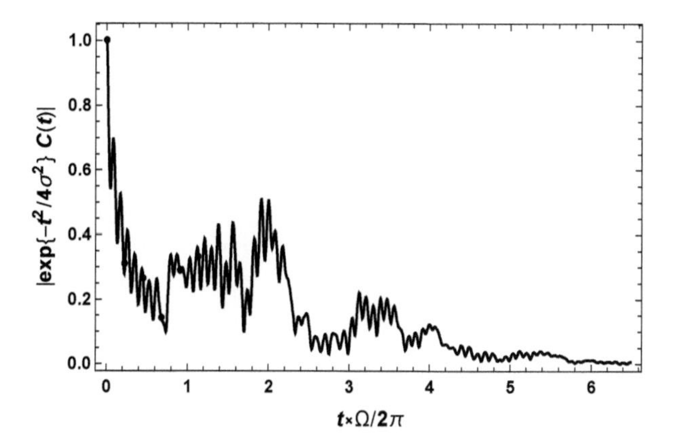

Fig. 3.3 The magnitude of the Heller overlap times the temporal window versus time in g-state slow-mode periods. Values taken at several small-integer multiples of 5/2 times the fast-mode vibrational period in the e-state are marked with dots.

is pushed to larger X values, with a high-frequency sawtooth structure resulting from small-amplitude oscillatory motion in the x coordinate. Subsequent broad recurrences arise from the wave packet's unfocused partial return to the Franck–Condon region, where it regains overlap with the overall vibrational ground state. Some values of the plotted quantity at intervals of $2.5 \times 2\pi/\omega_e$ are highlighted with black dots.

Figure 3.4 samples the wave-packet dynamics in the excited electronic state. The real part of the evolving two-dimensional e-state nuclear wave function is shown at the six equally spaced instants highlighted in Fig. 3.3. The overall phase is referenced to that of the lowest-lying vibronic level in the excited electronic state. The interval between the snapshots is a small fraction, $0.2273 \times 2\pi/\Omega$, of the slow mode's g-state period.

The time-evolving average values of X and x are displayed in Fig. 3.4. Both expectation values are seen to move away from their equilibrium values in the ground state and to oscillate about the e-state product well. Episodes of small $\langle X\rangle$ coincide with the times of relatively large overlap between ground- and excited-state wave packets observed in Fig. 3.3.

3.3.2 One-color transient-absorption signals

Here, we consider "one-color" transient-absorption signals resulting from the response of our model system to degenerate pump and probe pulses, both of which are resonant

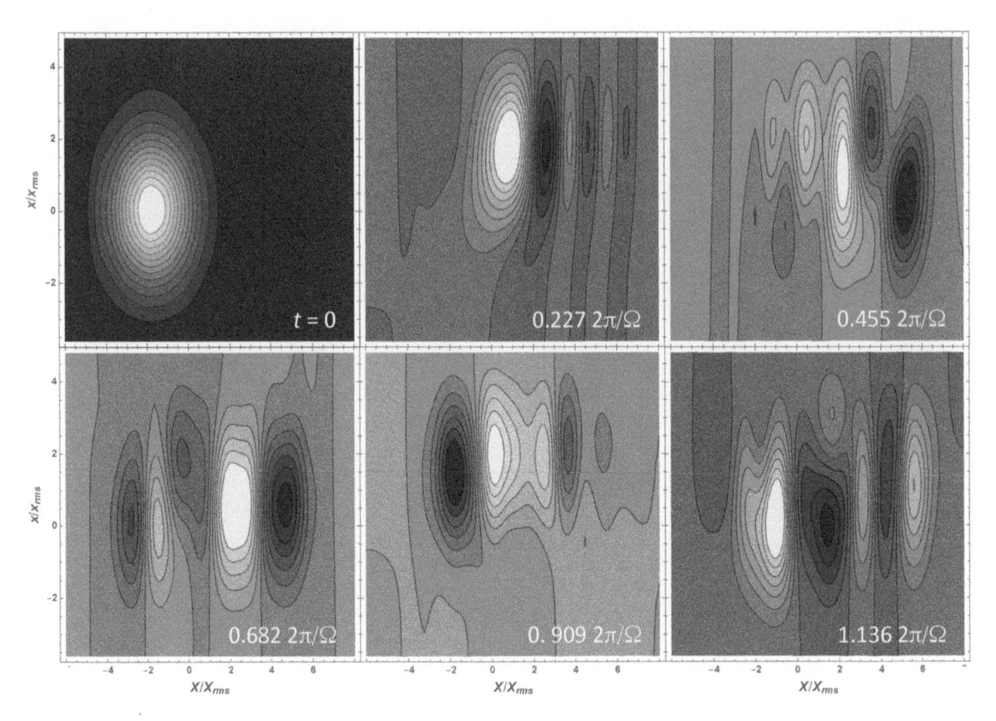

Fig. 3.4 Real part of nuclear wave packet in the excited electronic state at $t = 0, 1, ..., 5$ times $2.5\,(2\pi/\omega_e)$, the times marked in the plot of Heller's absorption kernel.

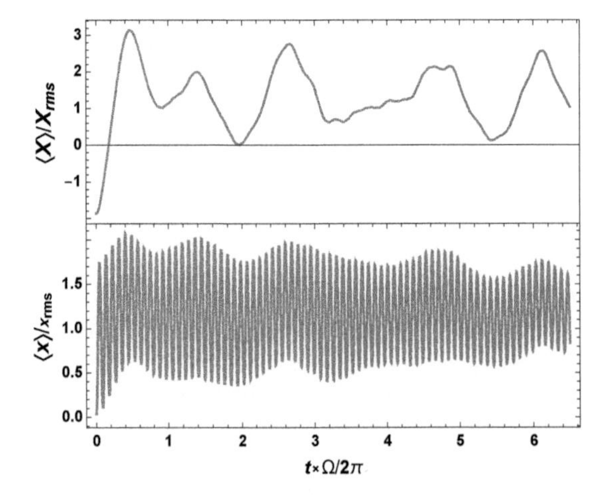

Fig. 3.5 $\langle X \rangle$ and $\langle x \rangle$ versus time during wave-packet motion in the e-state.

with the g-to-e electronic transition.[4] Because these pulses are *non*-resonant with $f \leftrightarrow e$, all of the overlaps in the transient-absorption dipole of eqn (3.18) involving amplitude transfer to or from the f-state become negligibly small, and it simplifies to

$$m_{u^2r}(t) \cong \langle \uparrow_u \downarrow_u \uparrow_r |\hat{m}|0\rangle + \langle \uparrow_r \downarrow_u \uparrow_u |\hat{m}|0\rangle + \langle \uparrow_r |\hat{m}| \uparrow_u \downarrow_u\rangle + \langle \uparrow_u |\hat{m}| \uparrow_u \downarrow_r\rangle + \text{c.c.} \quad (3.34)$$

According to the pattern established in Section 3.2.2, the first of these terms gives rise to a ground-state-bleach signal contribution,

$$\Delta\mathcal{E}_{\bar{\omega}\text{GSB}} = -\hbar\bar{\omega}\frac{\tilde{f}_r(\bar{\omega} - \Lambda_r)}{\sigma_r}\big(F_u^{(eg)}\big)^2\big(F_r^{(eg)}\big)^2\langle 0_g|p_u^{(ge)}(\bar{\tau};\bar{\bar{\tau}})p_u^{(eg)}(\tau + t_d;\bar{\tau})[-t_d]_{gg} \\ \times p_r^{(ge)}(t;\tau)p_{r\bar{\omega}}^{(eg)}(\infty;t)[t_d]_{gg}|0_g\rangle + \text{c.c.} \quad (3.35)$$

(compare eqn (3.28)). The second term in eqn (3.34) requires the probe to act before the pump and will be neglected here. The third term generates the impulsive Raman signal contribution,

$$\Delta\mathcal{E}_{\bar{\omega}\text{ISRS}} = -\hbar\bar{\omega}\frac{\tilde{f}_r(\bar{\omega} - \Lambda_r)}{\sigma_r}\big(F_u^{(eg)}\big)^2\big(F_r^{(eg)}\big)^2\langle 0_g|[-t_d]_{gg}\, p_r^{(ge)}(t;\bar{\bar{\tau}})p_{r\bar{\omega}}^{(eg)}(\infty;t) \\ \times [t_d]_{gg}\, p_u^{(ge)}(t + t_d;\tau)p_u^{(eg)}(\tau;\bar{\tau})|0_g\rangle + \text{c.c.} \quad (3.36)$$

The fourth term produces the stimulated-emission signal,

$$\Delta\mathcal{E}_{\bar{\omega}\text{SE}} = -\hbar\bar{\omega}\frac{\tilde{f}_r(\bar{\omega} - \Lambda_r)}{\sigma_r}\big(F_u^{(eg)}\big)^2\big(F_r^{(eg)}\big)^2\langle 0_g|p_u^{(ge)}(t + t_d;\bar{\bar{\tau}})[-t_d]_{ee}\, p_{r\bar{\omega}}^{(eg)}(\infty;t) \\ \times p_r^{(ge)}(t;\tau)[t_d]_{ee}\, p_u^{(eg)}(\tau + t_d;\bar{\tau})|0_g\rangle + \text{c.c.} \quad (3.37)$$

The minus in front of these signals corresponds to the fact that in each case the pump pulse tends to decrease probe absorption. In both GSB and ISRS contributions, wave-packet evolution during the interpulse delay occurs in the electronic ground state, whereas in SE it is in the excited state.

Adopting a simplification that affects the calculated signal only at delay times as short as the pulse durations, we neglect the truncation of pump action by the probe. In the stimulated-emission contribution, for instance, we make the replacement,

$$p_r^{(ge)}(t;\tau)[t_d]_{ee}\, p_u^{(eg)}(\tau + t_d;\bar{\tau}) \cong p_r^{(ge)}(t;\tau)[t_d]_{ee}\, p_u^{(eg)}(\infty;\bar{\tau})\,. \quad (3.38)$$

This approximation ceases to matter once t_d exceeds the pulse durations and the matrix elements of $p_u^{(eg)}$ take constant values (see below). Detailed attention must be paid to the implementation of this approximation in the case of a pump-pulse propagator whose upper limit involves the integration variable of a reduced propagator for the *spectrally filtered* probe pulse. For example, the impulsive stimulated Raman signal of eqn (3.36) includes the combination $p_r^{(ge)}(t;\bar{\bar{\tau}})p_{r\bar{\omega}}^{(eg)}(\infty;t)[t_d]_{gg}\, p_u^{(ge)}(t + t_d;\tau)$, in which

[4]We postpone to Chapter 5 the calculation and interpretation of two-color transient-absorption signals from a similar model system in which a pump-pulse resonant with g-to-e is accompanied by a probe-pulse resonant with e-to-f.

the pump is truncated by the integration variable of the (temporally elongated) filtered probe. The left-most term, $p_r^{(ge)}(t;\bar{\bar{\tau}})$, only takes non-negligible values once t approaches or exceeds t_d; and we are seeking an accurate treatment only for t_d significantly longer than the pulse durations. In this delay range, $t + t_d$ in the argument of $p_u^{(ge)}$ is always larger than σ_u, and we can write,

$$p_r^{(ge)}(t;\bar{\bar{\tau}})p_{r\bar{\omega}}^{(eg)}(\infty;t)[t_d]_{gg}\,p_u^{(ge)}(t+t_d;\tau) \cong p_r^{(ge)}(t;\bar{\bar{\tau}})p_{r\bar{\omega}}^{(eg)}(\infty;t)[t_d]_{gg}\,p_u^{(ge)}(\infty;\tau)\,. \tag{3.39}$$

Show that the same reasoning can be applied to $p_u^{(ge)}(t + t_d;\bar{\bar{\tau}})[-t_d]_{ee}p_{r\bar{\omega}}^{(eg)}(\infty;t) \times p_r^{(ge)}(t;\tau)$ appearing in eqn (3.37) for the stimulated-emission signal.

Under these approximations, the three relevant contributions (3.35)-(3.37) to the one-color transient-absorption signal take the simpler forms

$$\Delta\mathcal{E}_{\bar{\omega}\mathrm{GSB}} = -\hbar\bar{\omega}\frac{\tilde{f}_r(\bar{\omega}-\Lambda_r)}{\sigma_r}\big(F_u^{(eg)}\big)^2\big(F_r^{(eg)}\big)^2\langle 0_g|p_u^{(ge)}(\bar{\tau};\bar{\bar{\tau}})p_u^{(eg)}(\infty;\bar{\tau})[-t_d]_{gg}$$
$$\times\, p_r^{(ge)}(t;\tau)p_{r\bar{\omega}}^{(eg)}(\infty;t)[t_d]_{gg}|0_g\rangle + \text{c.c.}\,, \tag{3.40}$$

$$\Delta\mathcal{E}_{\bar{\omega}\mathrm{ISRS}} = -\hbar\bar{\omega}\frac{\tilde{f}_r(\bar{\omega}-\Lambda_r)}{\sigma_r}\big(F_u^{(eg)}\big)^2\big(F_r^{(eg)}\big)^2\langle 0_g|[-t_d]_{gg}\,p_r^{(ge)}(t;\bar{\bar{\tau}})p_{r\bar{\omega}}^{(eg)}(\infty;t)$$
$$\times\,[t_d]_{gg}\,p_u^{(ge)}(\infty;\tau)p_u^{(eg)}(\tau;\bar{\tau})|0_g\rangle + \text{c.c.}\,, \tag{3.41}$$

and

$$\Delta\mathcal{E}_{\bar{\omega}\mathrm{SE}} = -\hbar\bar{\omega}\frac{\tilde{f}_r(\bar{\omega}-\Lambda_r)}{\sigma_r}\big(F_u^{(eg)}\big)^2\big(F_r^{(eg)}\big)^2\langle 0_g|p_u^{(ge)}(\infty;\bar{\bar{\tau}})[-t_d]_{ee}\,p_{r\bar{\omega}}^{(eg)}(\infty;t)$$
$$\times\, p_r^{(ge)}(t;\tau)[t_d]_{ee}\,p_u^{(eg)}(\infty;\bar{\tau})|0_g\rangle + \text{c.c.}\,, \tag{3.42}$$

respectively.

With a choice of Gaussian pulses having envelopes $f_j(t) = \exp\{-t^2/2\sigma_j^2\}$ for $j = u$ or r, the relevant first-order reduced pulse propagators have forms like

$$\langle n_e|p_j^{(eg)}(t;\tau)|0_g\rangle = \langle n_e|0_g\rangle\int_{-\infty}^{t}\frac{d\tau}{\sigma_j}\,\exp\Big\{-\frac{\tau^2}{2\sigma_j^2}+\frac{i\tau}{\hbar}\big(\epsilon_e+\epsilon_{n_e}-\epsilon_{0_g}-\hbar\Lambda_j\big)\Big\}$$
$$= \langle n_e|0_g\rangle\sqrt{\tfrac{\pi}{2}}\,e^{-s^2/2}\big\{1+\mathrm{erf}\big[\tfrac{1}{\sqrt{2}}\big(\tfrac{t}{\sigma_j}-is\big)\big]\big\}\,, \tag{3.43}$$

where $s = \frac{\sigma_j}{\hbar}(\epsilon_e + \epsilon_{n_e} - \epsilon_{0_g} - \hbar\Lambda_j)$ (i.e., the dimensionless offset of $n_e \leftarrow 0_g$ from resonance) and $\mathrm{erf}(x) = \frac{2}{\sqrt{\pi}}\int_0^x dy\, e^{-y^2}$. This quantity is plotted in Fig. 3.6 with respect to its arguments s and t. Notice that the real part of the reduced propagator begins to take non-negligible values as the light pulse turns on and remains non-negligible thereafter, whereas the imaginary part is substantial only during the short episode of pulse action. Both portions become small for transitions displaced from the carrier frequency by more than $\sim 2\pi/\sigma_j$.

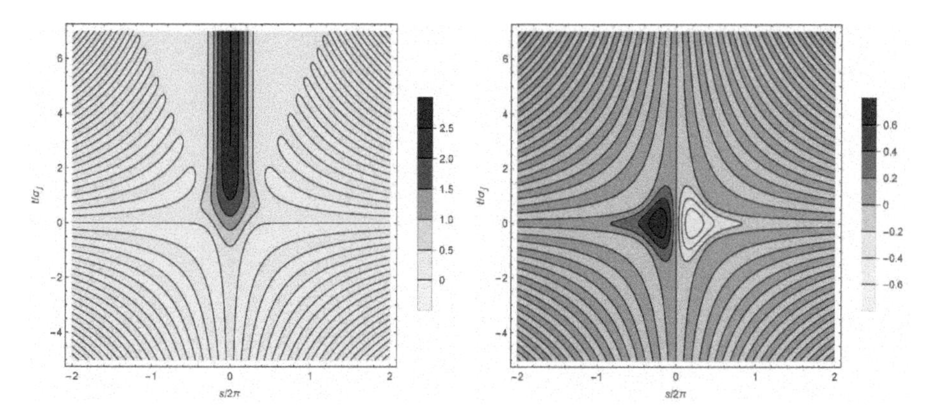

Fig. 3.6 Real part (left) and imaginary part (right) of $p_j^{(eg)}(t;\tau)$ divided by the Franck–Condon overlap $\langle n_e|0_g\rangle$, as functions of $s = (\epsilon_e + \epsilon_{n_e} - \epsilon_{0_g} - \hbar\Lambda_j)\sigma_j/\hbar$ and t/σ_j.

The ground-state bleach signal of eqn (3.40) and the impulsive Raman signal of eqn (3.41) both call for the nested double pump-pulse propagator, $p_u^{(ge)}(\infty;\tau)p_u^{(eg)}(\tau;\bar{\tau})$. Nuclear matrix elements of this operator are easily obtained by inserting a completeness relation $1 = \sum_{n_e}|n_e\rangle\langle n_e|$ and changing the variables of integration to $x = (\tau+\bar{\tau})/\sqrt{2}$ and $y = (\tau-\bar{\tau})/\sqrt{2}$. This procedure leads to

$$\langle n_g|p_u^{(ge)}(\infty;\tau)p_u^{(eg)}(\tau;\bar{\tau})|0_g\rangle = \pi e^{-\frac{\sigma_u^2}{4\hbar^2}(\epsilon_{n_g}-\epsilon_{0_g})^2}\sum_{n_e}\langle n_g|n_e\rangle\langle n_e|0_g\rangle e^{-\xi^2}[1+\mathrm{erf}(i\xi)], \tag{3.44}$$

with $\xi = [\hbar\Lambda_u + (\epsilon_{n_g}+\epsilon_{0g})/2 - \epsilon_e - \epsilon_{n_e}]\,\sigma_u/\hbar$. The summand in eqn (3.44) is a product of a term that depends only on the nuclear eigenenergies and terms that depend on the form of the corresponding eigenfunctions. The overall factor $e^{-\sigma_u^2(\epsilon_{n_g}-\epsilon_{0g})^2/4\hbar^2}$ enforces a requirement that the initial and final nuclear states lie within the spectral bandwidth of the pump pulse. The factor $e^{-\xi^2}[1+\mathrm{erf}(i\xi)]$ might appear similarly to suggest that $\epsilon_e + \epsilon_{n_e} - (\epsilon_{n_g} + \epsilon_{0g})/2$ must equal $\hbar\Lambda_u$ within $\pm 2\pi\hbar/\sigma_u$ in order for a given e-state vibrational level to contribute to the double pulse-propagator; that this is emphatically not the case is illustrated by Fig. 3.7, which plots the real and imaginary parts of this factor as a function of ξ. While the real part is just $e^{-\xi^2}$, the imaginary part is $e^{-\xi^2}\frac{2}{\sqrt{\pi}}\int_0^\xi dz\, e^{z^2}$, which is seen to take its values of largest magnitude at $\xi \cong \pm 0.924$ and to fall off much more slowly with increasing $|\xi|$ than the real part.[5] This behavior of the second-order pulse propagator accounts for the efficacy of

[5]For large positive ξ, we have $\frac{2}{\sqrt{\pi}}\int_0^\xi dz\, e^{z^2-\xi^2} \cong \frac{2}{\sqrt{\pi}}e^{-2\xi^2}\int_0^\xi dz\, e^{2\xi z} = \frac{1}{\sqrt{\pi}\xi}(1-e^{-2\xi^2})$, so the imaginary part of the double pulse-propagator falls off inversely as the resonance offset.

preresonant impulsive stimulated Raman excitation.[6,7]

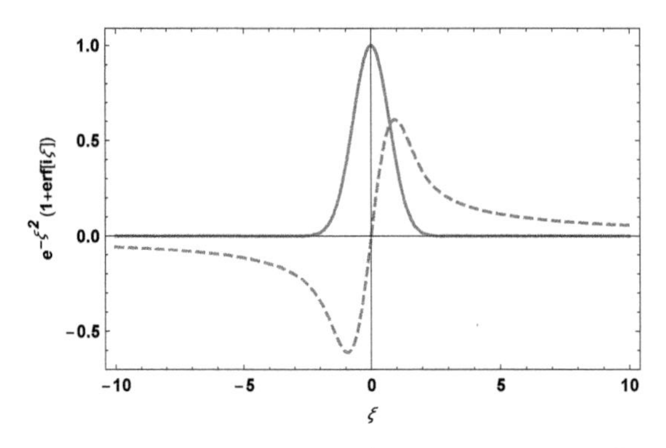

Fig. 3.7 Real (solid) and imaginary (dashed) parts of $e^{-\xi^2}[1 + \text{erf}(i\xi)]$.

Verify the derivations of (3.43) and (3.44).

Under the current neglect of pump-probe overlap effects, the remaining second-order pulse propagators are those involving the probe pulse nested inside the spectrally filtered probe, with matrix elements such as

$$\langle n_e | p_{r\bar{\omega}}^{(eg)}(\infty; t) p_r^{(ge)}(t; \tau) | \bar{n}_e \rangle = \sum_{n_g} \int_{-\infty}^{\infty} \frac{dt}{\bar{\sigma}} \frac{\sin \frac{\delta\omega}{2} t}{\frac{\delta\omega}{2} t} e^{-i\bar{\omega} t} \int_{-\infty}^{t} \frac{d\tau}{\sigma_r} e^{-\frac{\tau^2}{2\sigma_r^2} + i\Lambda_r \tau}$$
$$\times \langle n_e | [-t]_{ee} [t]_{gg} | n_g \rangle \langle n_g | [-\tau]_{gg} [\tau]_{ee} | \bar{n}_e \rangle \,. \tag{3.45}$$

The outer, t-integration takes the form of a windowed Fourier transformation; it could easily be carried out numerically in the present calculations. But the sinc-like form of its prolonged temporal window is just an artifact of our idealized treatment of spectral filtration.[8] It is therefore expedient—and little affects the outcome—to make the replacement

$$\frac{\sin \frac{\delta\omega}{2} t}{\frac{\delta\omega}{2} t} \cong \frac{2}{\delta\omega\, t} \left[\frac{\delta\omega}{2} t - \frac{1}{6} \left(\frac{\delta\omega}{2} t \right)^3 \right] \cong e^{-\frac{1}{6} \left(\frac{\delta\omega}{2} t \right)^2} , \tag{3.46}$$

[6] L. Dhar, J. A. Rogers, and K. A. Nelson, "Time-resolved vibrational spectroscopy in the impulsive limit," Chem. Rev. **94**, 157–193 (1994); T. J. Smith and J. A. Cina, "Toward preresonant impulsive Raman preparation of large amplitude vibrational motion," J. Chem. Phys. **104**, 1272–1292 (1996).

[7] In the limit of very short σ_u, both $e^{-\sigma_u^2(\epsilon_{n_g} - \epsilon_{0g})^2/4\hbar^2}$ and $e^{-\xi^2}[1 + \text{erf}(i\xi)]$ approach unity, and we have $\langle n_g | p_u^{(ge)}(\infty; \tau) p_u^{(eg)}(\tau; \bar{\tau}) | 0_g \rangle \cong \pi \sum_{n_e} \langle n_g | n_e \rangle \langle n_e | 0_g \rangle = \pi \langle n_g | 0_g \rangle$; vibronic resonance is ensured regardless of the precise value of Λ_u, and the second-order action of the pump pulse simply accounts for partial evacuation of the initially populated nuclear level.

[8] Radiative or non-radiative decay of excited states often occur on timescales competitive with $2\pi/\delta\omega$, and optical dephasing in multimode systems typically occurs more quickly still. These dynamical features diminish the importance of the precise form of the filtered-probe's temporal envelope.

which is correct through second order in t and vanishes for $|t| \to \infty$. With this substitution, it becomes possible to evaluate eqn (3.45) explicitly, yielding

$$\langle n_e | p_{r\bar{\omega}}^{(eg)}(\infty; t) p_r^{(ge)}(t; \tau) | \bar{n}_e \rangle = \sqrt{3} \sum_{n_g} \langle n_e | n_g \rangle \langle n_g | \bar{n}_e \rangle \exp\left\{ -\frac{6\lambda^2}{\delta\omega^2} - \frac{\bar{\lambda}^2 \sigma_r^2}{2} \right\}$$
$$\times \left\{ 1 + \text{erf}\left[\frac{i}{\delta\omega} \frac{12\lambda + \bar{\lambda}\sigma_r^2 \delta\omega^2}{\sqrt{24 + 2\sigma_r^2 \delta\omega^2}} \right] \right\}, \qquad (3.47)$$

where $\hbar\lambda = \epsilon_e + \epsilon_{n_e} - \epsilon_{n_g} - \hbar\bar{\omega}$ and $\hbar\bar{\lambda} = \epsilon_e + \epsilon_{\bar{n}_e} - \epsilon_{n_g} - \hbar\Lambda_r$. The summand in eqn (3.47) is a product of contributing Franck–Condon overlaps with a factor that depends on the offset of the corresponding transition frequencies from $\bar{\omega}$ and Λ_r. We now have at our disposal representative examples of matrix elements of all the reduced pulse-propagator combinations needed to calculate transient-absorption signals from our model system.

1. Derive eqn (3.47) from eqns (3.45) and (3.46) by making successive variable changes to $(\bar{t} = t\delta\omega/2\sqrt{6}, \bar{\tau} = \tau/\sigma_r\sqrt{2})$, then $(U = \bar{t}\sin\theta + \bar{\tau}\cos\theta, u = \bar{t}\cos\theta - \bar{\tau}\sin\theta)$, with $\cos\theta = 2\sqrt{3}/\sqrt{12 + \sigma_r^2\delta\omega^2}$ and $\sin\theta = \sigma_r\delta\omega/\sqrt{12 + \sigma_r^2\delta\omega^2}$. As a check on your derivation (and mine) you may consider the unrealistic situation in which $\delta\omega = 2\sqrt{3}/\sigma_r$ and $\bar{\omega} = \Lambda_r$, and compare the result to eqn (3.44).
2. Obtain an approximate formula for $\langle n_g | p_{r\bar{\omega}}^{(ge)}(\infty; t) p_r^{(eg)}(t; \tau) | \bar{n}_g \rangle$, which is needed for the GSB and ISRS contributions, by analogy with eqn (3.47).

Now we can take a look at the one-color transient-absorption spectrum of our model system. The components given by eqns (3.40)–(3.42) were calculated with identical pump and probe pulses having center frequencies $\Lambda_u = \Lambda_r$ matching $\epsilon_e/\hbar = 600\,\Omega$ and durations $\sigma_u = \sigma_r = 0.20(2\pi/\omega_e)$. The dectector slit-width is set to $\delta\omega = 3.33\,\Omega$, so that $\bar{\sigma} = 250$ fs remains sufficiently short to resolve motion on the 833-fs timescale of the 40-cm^{-1} mode.

The calculated contributions along with the net transient-absorption signal are shown in Figs 3.8 and 3.9. Each of these mostly negative quantities is plotted as the hyperbolic tangent of the corresponding component divided by about 87% of its largest magnitude in order to accentuate the lower-amplitude features.[9] Those maximum magnitudes take relative proportions $1:1.07:1.98:3.70$ for the ground-state bleach, impulsive stimulated Raman, stimulated emission, and total signals, respectively. Interpulse delay is reckoned in periods of the slow ground-state vibration, and the exhibited spectral range is nearly twice the full width at half maximum of the pulses' power spectrum.

The central stripes in the GSB and ISRS contributions undulate at frequency Ω, while sideband oscillations occur at the higher frequency ω_g. In the stimulated-emission portion, e-state dynamics at the round-trip time of the slower motion and the vibrational frequency ω_e both become manifest. Because the pump pulse is *much* shorter than $2\pi/\Omega$, impulsive Raman excitation of the slower mode becomes inefficient,

[9] The plotted quantities are independent of E_u, E_r, and the transition dipole moment m_{eg}, so we need not specify values for these quantities.

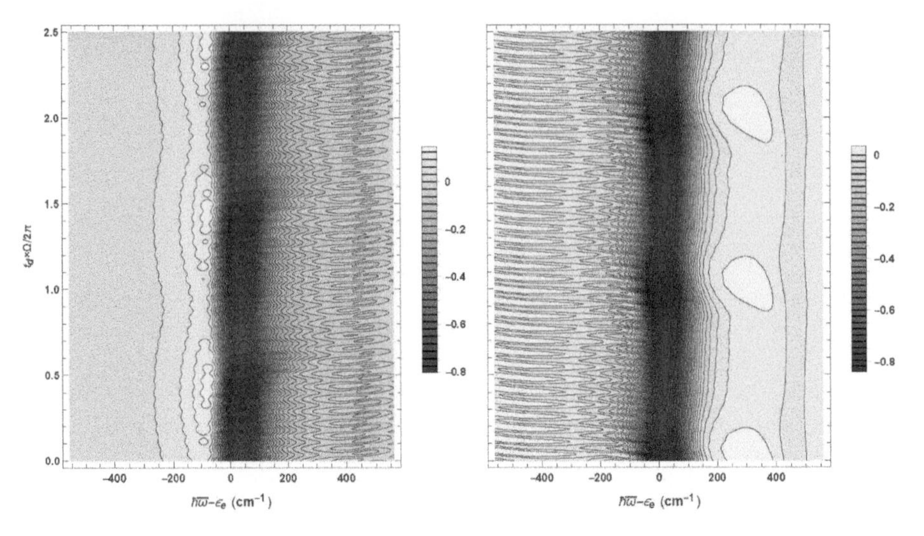

Fig. 3.8 Ground-state bleach (left) and impulsive stimulated Raman (right) components of the transient-absorption signal from the two-dimensional model system with degenerate, electronically resonant pump and probe pulses. Hyperbolic-tangent scaling is used here and in the next figure to spread the most closely spaced contours.

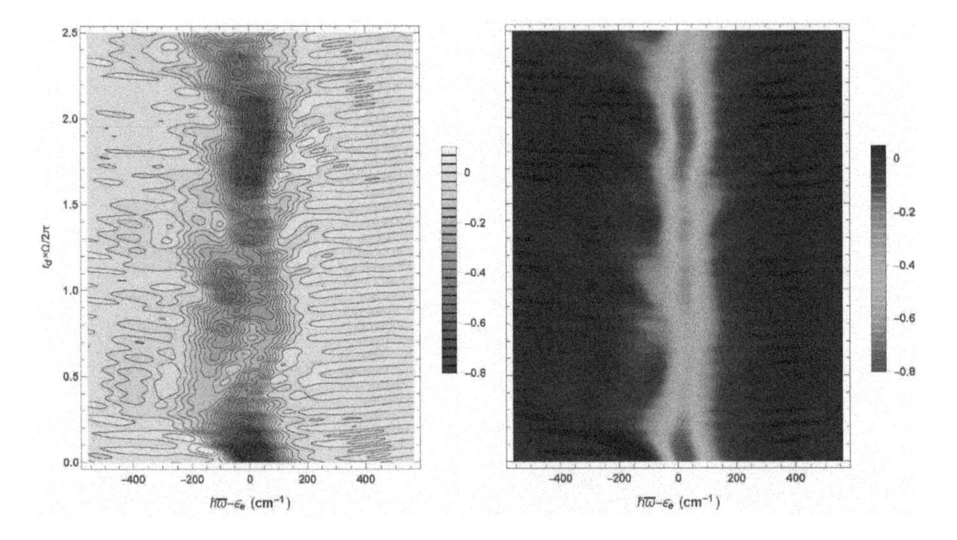

Fig. 3.9 Stimulated emission (left) and total signal (right) from the 2D model.

and the low-frequency features in GSB and ISRS are less prominent than the Franck–Condon displacement-driven slow motion in the SE signal. The blue sidebands in GSB and the red ones in ISRS are seen to be out of phase with each other. While sideband structure is present on both blue and red sides of the SE spectrum, the latter appears more complicated. The two sides of the SE signal more obviously exhibit anti-phasing with the use of intensity scaling that gives still greater emphasis to low-amplitude features (not shown). The experimentally observable total spectrum superposes all three contributions.

Detailed analysis of the TA signal is possible, based on the temporal profile along each spectral slice.[10,11] The intensity of the signal at a given t_d and $\bar{\omega}$ can be qualitatively related to the probability amplitude of the relevant g- or e-state nuclear wave packet at that time along the locus of coordinates having that particular value of the local electronic transition frequency. This feature accounts for the anti-phasing that is sometimes observed between red- and blue-side quantum beats.[12]

Photo-induced dynamics in an excited electronic state is often explored using two-color transient-absorption measurements in which the probe pulse is tuned to match an excited-state-to-excited-state electronic transition akin to the $f \leftarrow e$ transition in the model system studied in this chapter. Although the signal calculations carried out here are restricted to one-color measurements to which the latter transition does not contribute, Chapter 5 revisits transient-absorption spectroscopy and presents calculated signals for the complementary (two-color) case of a similar molecular model.

[10]W. T. Pollard, S. L. Dexheimer, Q. Wang, L. A. Peteanu, C. V. Shank, and R. A. Mathies, "Theory of dynamic absorption spectroscopy of nonstationary states. 4. Application to 12-fs resonant impulsive Raman spectroscopy of bacteriorhodopsin," J. Phys. Chem. **96**, 6147–6158 (1992).

[11]J. Brazard, L. A. Bizimana, T. Gellen, W. P. Carbery, and D. B. Turner, "Experimental detection of branching at a conical intersection in a highly fluorescent molecule," J. Phys. Chem. Lett. **7**, 14–19 (2016).

[12]J. A. Cina *et al.* (2016), cited on page 18.

4

How fissors works: Femtosecond stimulated Raman spectroscopy as a probe of conformational change

4.1 Basic idea

Femtosecond stimulated Raman spectroscopy (FSRS or fissors) is a form of time-resolved Raman-scattering spectroscopy designed to probe changes in molecular conformation. Because the energy-level splittings associated with conformational motion are small, in keeping with the relative slowness of changes in overall molecular shape, the correspondingly small Raman shifts, frequency changes associated with these transitions, can be difficult to resolve against the background of a strong, unshifted Rayleigh-scattering signal. As an alternative, FSRS monitors the time-varying spectral location and shape of the more dramatically shifted Raman signals associated with higher frequency vibrations and tracks conformational change through the anharmonic coupling of those vibrations to the conformational modes.[1,2]

In order to bring out the basic notions of signal generation and information content in a fissors measurement, we consider a molecule with the Hamiltonian,

$$H = |g\rangle H_g \langle g| + |e\rangle H_e \langle e| + |f\rangle H_f \langle f| \,, \tag{4.1}$$

which comprises three electronic states, labeled g, e, and f, and just two nuclear degrees of freedom. The nuclear Hamiltonian of the jth electronic state can be written

$$H_j = \frac{P^2}{2M} + h_j(Q) \,, \tag{4.2}$$

in which, anticipating a vibrationally adiabatic approximation, we have identified vibrational Hamiltonians,

$$h_j(Q) = \frac{p^2}{2m} + \frac{m\omega^2(Q)}{2}\big(q - q_j(Q)\big)^2 + v_j(Q) \,, \tag{4.3}$$

[1]The scope of FSRS applications is sampled in D. R. Dietze and R. A. Mathies, "Femtosecond stimulated Raman spectroscopy," ChemPhysChem **17**, 1224–1251 (2016). For a theoretical description, see S.-Y. Lee, D. Zhang, D. W. McCamant, P. Kukura, and R. A. Mathies, "Theory of femtosecond stimulated Raman spectroscopy," J. Chem. Phys. **121**, 3632–3642 (2004).

[2]Ultrafast Raman spectroscopy more broadly is surveyed by H. Kuramochi and T. Tahara, "Tracking ultrafast structural dynamics by time-domain Raman spectroscopy," J. Am. Chem. Soc. **143**, 9699–9717 (2021).

parametrized by the conformational coordinate operator Q. The vibrational frequency $\omega(Q)$ is taken to be the same function of Q in all three electronic states, and the vibrational period $\sim 2\pi/\omega(Q)$ is much shorter than the timescale of conformational motion. The equilibrium g-state vibrational coordinate $q_g(Q)$ is set to zero, while a displacement between e- and f-state minima, $q_f(Q) = q_e(Q) + \delta q$, makes the $f \leftrightarrow e$ transition vibrationally Raman active. The conformational potentials $v_g(Q)$ and $v_e(Q)$ are assumed to differ by more than a constant electronic energy shift, so that electronic excitation by the launch pulse will trigger conformational change, but we simply let $v_f(Q) = v_e(Q) + \hbar\epsilon_{fe}$.

The eigenstates and eigenenergies of the vibrational Hamiltonians (4.3) obey

$$h_j(Q)|n_j(Q)\rangle = \hbar\epsilon_{n_j}(Q)|n_j(Q)\rangle \, , \tag{4.4}$$

where

$$\hbar\epsilon_{n_j}(Q) = \hbar\omega(Q)(n_j + \tfrac{1}{2}) + v_j(Q) \, . \tag{4.5}$$

Consistently with the assumption that the vibrational state follows the molecular conformation adiabatically, the nuclear eigenstates are taken to have the form $|n_j(Q)\rangle|m_{n_j}\rangle$, in which, significantly, the conformational coordinate operator Q parametrizing the vibrational eigenstate operates on the conformational state. These states satisfy the nuclear Schrödinger equation,

$$\begin{aligned} H_j|n_j(Q)\rangle|m_{n_j}\rangle &= \Big(\frac{P^2}{2M} + h_j(Q)\Big)|n_j(Q)\rangle|m_{n_j}\rangle \\ &= \Big(\frac{P^2}{2M} + \hbar\epsilon_{n_j}(Q)\Big)|n_j(Q)\rangle|m_{n_j}\rangle \\ &\cong |n_j(Q)\rangle\Big(\frac{P^2}{2M} + \hbar\epsilon_{n_j}(Q)\Big)|m_{n_j}\rangle \\ &= |n_j(Q)\rangle|m_{n_j}\rangle\hbar\mathcal{E}_{m_{n_j}} \, . \end{aligned} \tag{4.6}$$

For any choice of electronic and vibrational indices, j and n, the conformational energy eigenkets $\{|m_{n_j}\rangle\}$ provide a basis set for the conformational degrees of freedom. Subsequent sections consider time-dependent wave packets capturing the conformational motion triggered by photo-excitation. In the numerical illustrations of Section 4.4, each relevant conformational wave packet is expanded across such a basis set.

The molecule interacts with three laser pulses under the time-dependent Hamiltonian, $H(t) = H - \hat{\mu}E(t)$, where

$$\hat{\mu} = \mu_{eg}\big(|e\rangle\langle g| + |g\rangle\langle e|\big) + \mu_{fe}\big(|f\rangle\langle e| + |e\rangle\langle f|\big) \, , \tag{4.7}$$

and $E(t) = E_0(t) + E_1(t) + E_2(t)$, with

$$E_I(t) = E_I f_I(t - t_I)\cos[\Omega_I(t - t_I) + \varphi_I] \, . \tag{4.8}$$

The optical phases φ_I (and their differences) are uncontrolled in fissors. We set the arrival time of the launch pulse ($I = 0$) to $t_0 = 0$, while the Raman pump pulse (1) and the Raman probe (2) arrive simultaneously at $t_1 = t_2 \equiv t_d$. The center frequency of the launch pulse, with $\hbar\Omega_0 \approx \upsilon_e - \upsilon_g$, is resonant with $e \leftarrow g$. The Raman pump and probe are both *sub*-resonant with the lower-frequency $f \leftrightarrow e$ transition, $\hbar\Omega_2 < \hbar\Omega_1 < \upsilon_f - \upsilon_e < \upsilon_e - \upsilon_g$, and $\Omega_1 - \Omega_2 \approx \omega$ so that these temporally coincident pulses can drive vibrational Raman transitions in the electronic e-state. The Raman pump is lengthy on the vibrational timescale, and its approximate duration, $\tau_1 \gg 2\pi/\omega$, may even span the relevant time-range of conformational change. The launch and Raman probe pulses are conformationally abrupt, but their durations, $\tau_0, \tau_2 > 2\pi/\omega$, are slightly longer than the vibrational period, so that neither can independently generate vibrational coherence. See Fig. 4.1.

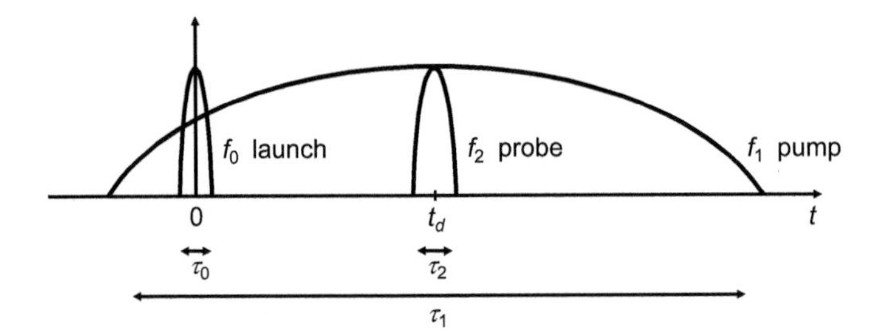

Fig. 4.1 Envelope functions for the launch (0), Raman pump (1), and Raman probe (2) laser pulses in femtosecond stimulated Raman spectroscopy. The launch pulse as well as the probe pulse may be embedded within the nearly monochromatic Raman pump, whose duration can be similar to the timescale of molecular conformational change. But the treatment developed here applies without modification in the simpler case when the launch pulse precedes the advent of the Raman pump pulse.

4.2 Signal formation

The induced dipole moment giving rise to the fissors field is of order $E_0^2 E_1^2 E_2$ in the incident field strengths. The fissors field interferes with the transmitted Raman probe field and both are spectrally resolved to produce a signal of order $E_0^2 E_1^2 E_2^2$ that is trilinear in the intensities of the launch, pump, and probe beams. In order to calculate the contributions of order $E_0^2 E_1^2 E_2$ to $\langle\Psi|\hat{\mu}|\Psi\rangle$, we need

$$|\Psi\rangle = |\uparrow_0\rangle + |\uparrow_0\uparrow_1\rangle + |\uparrow_0\uparrow_2\rangle + \underbrace{|\uparrow_0\uparrow_1\downarrow_1\rangle}_{\text{no Raman gain}} + |\uparrow_0\uparrow_1\downarrow_2\rangle$$
$$+ \underbrace{|\uparrow_0\uparrow_2\downarrow_1\rangle}_{\text{no gain}} + \underbrace{|\uparrow_0\uparrow_1\downarrow_1\uparrow_2\rangle}_{\text{no gain}} + \underbrace{|\uparrow_0\uparrow_1\downarrow_2\uparrow_1\rangle}_{\sim\exp\{-2i\varphi_1\}} + \underbrace{|\uparrow_0\uparrow_2\downarrow_1\uparrow_1\rangle}_{\text{no gain}} . \tag{4.9}$$

In each contributing multi-pulse ket of eqn (4.9), the order of successive action by the pulses on the system is listed from left to right. In $|\uparrow_0\uparrow_1\downarrow_2\rangle$ for instance, the molecule is excited from the g-state to the e-state by the launch pulse, then excited (subresonantly) to the f-state by the Raman pump pulse and returned to e under the contemporaneous influence of the Raman probe. As indicated, several of the kets in eqn (4.9) are unable, due to the frequency composition of the pulses, to effect vibrational transitions in the electronic e-state. In addition, the amplitude $|\uparrow_0\uparrow_1\downarrow_2\uparrow_1\rangle$ involves two upward transitions driven by the Raman pump; since it therefore carries an uncontrolled optical phase-factor $\exp\{-2i\varphi_1\}$, its contribution to any expectation value of the required order in the fields averages effectively to zero over the many laser shots needed to acquire a signal. From the remaining kets on the right-hand side of eqn (4.9), we see that the fissors dipole reduces to a single term,

$$\mu_{\rm FSRS} = 2{\rm Re}\{\langle\uparrow_0\uparrow_1|\hat{\mu}|\uparrow_0\uparrow_1\downarrow_2\rangle\} = 2\mu_{fe}{\rm Re}\{\langle\uparrow_0\uparrow_1|f\rangle\langle e|\uparrow_0\uparrow_1\downarrow_2\rangle\}\,; \tag{4.10}$$

the rightmost expression gives the fissors dipole as the overlap between a three-pulse nuclear wave packet in the e-state and a two-pulse wave packet in the f-state.

The induced dipole moment (4.10) radiates an electric field that is given in the far-field region by

$$E_{FSRS}(\mathbf{R},t) \cong \frac{\Omega_2^2}{c^2R}\,\mu_{\rm FSRS}\Big(t-\frac{R}{c}\Big)\,. \tag{4.11}$$

The fissors field (4.11) interferes with the electric field of the transmitted Raman probe pulse and both are spectrally filtered by a detector to produce the FSRS signal. This signal is a function the "monochromator" setting $\bar{\omega}$ (as well as the delay-time t_d). It can be expressed as the change in electromagnetic energy of the probe-plus-fissors field due to the presence of the Raman pump pulse:

$$\begin{aligned}\Delta\mathcal{U}(\bar{\omega}) &= \frac{1}{4\pi}\int d^3R\,\big\{\big[E_2'(\mathbf{R},t)+E_{FSRS}'(\mathbf{R},t)\big]^2 - E_2'^2(\mathbf{R},t)\big\}\\ &\cong \frac{1}{2\pi}\int d^3R\,E_2'(\mathbf{R},t)E_{FSRS}'(\mathbf{R},t)\,;\end{aligned} \tag{4.12}$$

the primes denote spectral filtration. With the Raman probe taken to propagate in the laboratory Z-direction, $E_2(\mathbf{R},t) = E_2f_2(t-t_2(\mathbf{R}))\cos[\Omega_2(t-t_2(\mathbf{R}))+\varphi_2]$, where $t_2(\mathbf{R}) = t_d + Z/c$ (compare eqn (4.8)). By Fourier transforming $E_2(\mathbf{R},t)$ to find its frequency components, windowing these between $\bar{\omega}-\delta\omega/2$ and $\bar{\omega}+\delta\omega/2$ (or $-\bar{\omega}-\delta\omega/2$ and $-\bar{\omega}+\delta\omega/2$), and inverse Fourier transforming, we can construct a carrier-wave/envelope expression

$$E_2'(\mathbf{R},t) = E_2'f_2'(t-t_2(\mathbf{R}))\cos[\bar{\omega}(t-t_2(\mathbf{R}))+\varphi_2]\,, \tag{4.13}$$

where $E_2' = E_2\frac{\delta\omega}{2\pi}\tilde{f}_2(\bar{\omega}-\Omega_2) = E_2\frac{\delta\omega}{2\pi}\int_{-\infty}^{\infty}d\tau\,e^{i\tau(\bar{\omega}-\Omega_2)}f_2(\tau)$ (assumed to be real) and $f_2'(t) = \frac{2}{\delta\omega\,t}\sin\frac{\delta\omega}{2}t$.

> 1. Carry out the exercise that leads to eqn (4.13).
> 2. The fundamental signal expression of eqn (4.12) becomes independent of time once t is large enough that the spatial range of the filtered probe field ceases to encompass the molecule's location at the origin. Discuss whether and by how much the time considered must be further increased so that the far-field condition $Z > 2\pi c/\Omega_2$ invoked in eqn (4.11) applies throughout the range of the filtered probe.

Since the filtered probe field (eqn (4.13)) comprises components with magnitudes of the frequency between $\bar{\omega}-\delta\omega/2$ and $\bar{\omega}+\delta\omega/2$, it comprises wave-vector components in the Z-direction with magnitudes between $\bar{\omega}/c-\delta\omega/2c$ and $\bar{\omega}/c+\delta\omega/2c$. The spatial integration in eqn (4.12) would then select for the wave-vector components of $E'_{FSRS}(\mathbf{R},t)$ within the same range and hence for the corresponding frequency components. As a result, explicit use of the spectrally filtered fissors field becomes superfluous (its prime can be dropped) and the signal expression (4.12) can be replaced by

$$\Delta\mathcal{U}(\bar{\omega}) = \frac{1}{2\pi}\int d^3R\, E'_2(\mathbf{R},t)E_{FSRS}(\mathbf{R},t)\,. \tag{4.14}$$

This three-dimensional spatial integral can be further reduced to a one-dimensional integral over time. Written in terms of spherical polar coordinates, eqn (4.14) becomes

$$\begin{aligned}\Delta\mathcal{U}(\bar{\omega}) = \frac{E'_2}{2\pi}\int_0^\infty R\,dR\int_0^\pi \sin\theta\,d\theta\int_0^{2\pi} d\phi\, f'_2\big(t-t_d-\tfrac{R}{c}\cos\theta\big)\\ \times\cos\big[\bar{\omega}\big(t-t_d-\tfrac{R}{c}\cos\theta\big)+\varphi_2\big]\frac{\Omega_2^2}{c^2}\,\mu_{FSRS}\big(t-\tfrac{R}{c}\big)\,. \end{aligned}\tag{4.15}$$

The integral over the azimuthal angle cancels the 2π in the denominator out front. Integration by parts over the polar angle, together with recognition of the (very) slowly varying nature of the filtered-probe envelope function then leads to

$$\begin{aligned}\Delta\mathcal{U}(\bar{\omega}) &= -E'_2\frac{\Omega_2}{c}\int_0^\infty dR\, f'_2\big(t-t_d-\tfrac{R}{c}\big)\sin\big[\bar{\omega}\big(t-t_d-\tfrac{R}{c}\big)+\varphi_2\big]\mu_{FSRS}\big(t-\tfrac{R}{c}\big)\\ &= -E'_2\Omega_2\int_{-\infty}^{\infty} d\tau\, f'_2(\tau-t_d)\sin[\bar{\omega}(\tau-t_d)+\varphi_2]\,\mu_{FSRS}(\tau)\,. \end{aligned}\tag{4.16}$$

Derive eqn (4.16) from eqn (4.15), paying particular attention to the limits of integration in the final expression.

4.3 Fissors dipole

In this section, we derive explicit expressions for the two-pulse and three-pulse kets which determine the fissors dipole of eqn (4.10). To pave the way for an *electronically* adiabatic treatment of the Raman pump's very slowly varying envelope, we make a rotating-frame transformation, $|\Psi_R(t)\rangle = R(t)|\Psi(t)\rangle$, with $R(t) = |g\rangle\langle g| + |e\rangle\langle e| + |f\rangle e^{i\Omega_1(t-t_d)}\langle f|$. This transformation replaces the time-dependent Schrödinger equation, $|\dot{\Psi}\rangle = -\frac{i}{\hbar}H(t)|\Psi\rangle$, whose initial condition we take to be $|\Psi(-\tau_1)\rangle = |g\rangle|0_g(Q)\rangle|0_{0_g}\rangle e^{i\tau_1\mathcal{E}_{0_{0_g}}}$, with the evolution equation

$$\begin{aligned}|\dot{\Psi}_R(t)\rangle &= R(t)\big[i\Omega_1|f\rangle\langle f| - \tfrac{i}{\hbar}H + \tfrac{i}{\hbar}\hat{\mu}E(t)\big]|\Psi(t)\rangle\\ &= -\frac{i}{\hbar}\big[H_R(f_1(t-t_d)) + V_0(t) + V_2(t)\big]|\Psi_R(t)\rangle\,, \end{aligned}\tag{4.17}$$

whose unaltered initial state is $|\Psi_R(-\tau_1)\rangle = |\Psi(-\tau_1)\rangle$. After making rotating-wave approximations in their definitions and ignoring far off-resonance terms in V_0 which couple the e- and f-states, along with those in V_1 and V_2 that couple g and e, we have

$$H_R(f_1) = |g\rangle H_g\langle g| + |e\rangle H_e\langle e| + |f\rangle(H_f - \hbar\Omega_1)\langle f| + V_1(f_1)\,, \tag{4.18}$$

with

$$V_1(f_1) = -\frac{\mu_{fe}E_1}{2} f_1\big(|f\rangle\langle e|e^{-i\varphi_1} + |e\rangle\langle f|e^{i\varphi_1}\big)\,, \tag{4.19}$$

$$V_0(t) = -\frac{\mu_{eg}E_0}{2} f_0(t)|e\rangle\langle g|e^{-i\varphi_0 - i\Omega_0 t} + \text{H.c.}\,, \tag{4.20}$$

$$V_2(t) = -\frac{\mu_{fe}E_2}{2} f_2(t - t_d)|e\rangle\langle f|e^{i\varphi_2 + i(\Omega_2 - \Omega_1)(t - t_d)} + \text{H.c.} \tag{4.21}$$

Note that V_1, which will be treated with an electronically adiabatic approximation, is taken to be a function of the Raman-pump envelope, f_1, whereas V_0 and V_2 are regarded as functions of time.[3]

Verify eqns (4.17) through (4.21).

We need the electronic eigenkets and eigenenergies of H_R only through first order in E_1. How come? These are found to be $|g\rangle$ with "eigenenergy" H_g,

$$|\xi(f_1)\rangle = |e\rangle - |f\rangle\frac{\theta}{2}e^{-i\varphi_1}\,, \tag{4.22}$$

with $\hbar\epsilon_\xi(f_1) = H_e$, and

$$|\xi'(f_1)\rangle = |e\rangle\frac{\theta}{2} + |f\rangle e^{-i\varphi_1}\,, \tag{4.23}$$

with $\hbar\epsilon_{\xi'}(f_1) = H_f - \hbar\Omega_1$. Here,

$$\theta \equiv -\frac{\mu_{fe}E_1}{H_f - \hbar\Omega_1 - H_e} f_1(t - t_d)\,; \tag{4.24}$$

because the kinetic energy operators cancel in the denominator, this quantity depends on nuclear coordinate operators alone.

Derive the formulas (4.22) and (4.23) for the eigenstates of $H_R(f_1)$, and confirm that their respective "eigenvalues" are as stated through first order in E_1.

We can re-express $H_R(f_1)$ in its approximate eigenbasis as

$$H_R(f_1) = |g\rangle H_g\langle g| + |\xi(f_1)\rangle H_e\langle\xi(f_1)| + |\xi'(f_1)\rangle(H_f - \hbar\Omega_1)\langle\xi'(f_1)|\,. \tag{4.25}$$

The launch and probe-pulse perturbations then become

$$V_0(t) \cong -\frac{\mu_{eg}E_0}{2} f_0(t)e^{-i\varphi_0 - i\Omega_0 t}\Big[\,|\xi\rangle\langle g| + \underbrace{|\xi'\rangle\langle g|\frac{\theta}{2}}_{\text{nonresonant}}\,\Big] + \text{H.c.}\,, \tag{4.26}$$

and

[3] Some curious readers pressed for time may wish simply to skim through the details of the following derivation and arrive at the final expression, eqn (4.50), for the fissors dipole.

$$V_2(t) \cong -\frac{\mu_{fe}E_2}{2} f_2(t-t_d) e^{i\varphi_2 - i\varphi_1 + i(\Omega_2-\Omega_1)(t-t_d)}$$
$$\times \Big[\underbrace{|\xi\rangle\langle\xi'|}_{\text{nonresonant}} + \frac{\theta}{2}\big(|\xi'\rangle\langle\xi'| - |\xi\rangle\langle\xi|\big)\Big] + \text{H.c.}, \tag{4.27}$$

respectively. As indicated, the $|\xi'\rangle\langle g|$ term in V_0 can be neglected because $|H_f - \hbar\Omega_1 - \hbar\Omega_0| > 2\pi\hbar/\tau_0$. Similarly, the $|\xi\rangle\langle\xi'|$ term in V_2 can be omitted since $|H_f - H_e - \hbar\Omega_2| > 2\pi\hbar/\tau_2$.

Do you find these arguments persuasive?

The remaining term in V_0, proportional to $|\xi\rangle\langle g|$, will be responsible for generating amplitude in the f-state via the $-|f\rangle(\theta/2)e^{-i\varphi_1}$ portion of $|\xi\rangle$ in eqn (4.22); operating perturbatively at first-order, V_0 therefore gives rise to $|\uparrow_0\uparrow_1\rangle$. There is also a second-order perturbation of the molecular state arising from the successive action of V_0 and V_2. Within this second-order term, there is a contribution in which V_0 first generates e-state amplitude via the $|e\rangle\langle g|$ portion of $|\xi\rangle\langle g|$ and the $\theta|\xi\rangle\langle\xi| \cong \theta|e\rangle\langle e|$ part of V_2 then prepares (vibrationally excited) amplitude in the same electronic state. This second-order perturbation is responsible for generating the three-pulse amplitude $|\uparrow_0\uparrow_1\downarrow_2\rangle$.[4]

In the electronically adiabatic basis, the molecular state can be written as $|\Psi_R\rangle = |g\rangle|\psi_g\rangle + |\xi\rangle|\psi_\xi\rangle + |\xi'\rangle|\psi_{\xi'}\rangle$, whose time derivative is

$$|\dot{\Psi}_R\rangle = |g\rangle|\dot{\psi}_g\rangle + |\dot{\xi}\rangle|\psi_\xi\rangle + |\xi\rangle|\dot{\psi}_\xi\rangle + |\dot{\xi}'\rangle|\psi_{\xi'}\rangle + |\xi'\rangle|\dot{\psi}_{\xi'}\rangle\,. \tag{4.28}$$

Now $|\dot{\xi}\rangle = |\xi\rangle\langle\xi|\dot{\xi}\rangle + |\xi'\rangle\langle\xi'|\dot{\xi}\rangle \cong |\xi\rangle\langle\xi|\dot{\xi}\rangle$, under the adiabatic condition where the pump-pulse envelope varies on a timescale much longer than the times $2\pi/|\epsilon_{\xi'} - \epsilon_\xi|$ associated with the ξ'-ξ electronic splitting. The inner product $\langle\xi|\dot{\xi}\rangle$ must be purely imaginary, as $\text{Re}\langle\xi|\dot{\xi}\rangle = \frac{1}{2}(\langle\xi|\dot{\xi}\rangle + \langle\dot{\xi}|\xi\rangle) = \frac{1}{2}\frac{d}{dt}\langle\xi|\xi\rangle = 0$. While the expansion coefficients of $|\xi\rangle$ in terms of $|e\rangle$ and $|f\rangle$ may be complex, those of $|\bar{\xi}\rangle \equiv W|\xi\rangle = c_e|e\rangle + c_f|f\rangle$, where $W = |e\rangle\langle e| + |f\rangle e^{i\varphi_1}\langle f|$, are real; and the inner product $\langle\bar{\xi}|\dot{\bar{\xi}}\rangle = c_e\dot{c}_e + c_f\dot{c}_f$ is real as well. Since $\langle\xi|\dot{\xi}\rangle = \langle\xi|W^\dagger W|\dot{\xi}\rangle = \langle\bar{\xi}|\dot{\bar{\xi}}\rangle$ must therefore be both real and imaginary, it has to vanish. Under the electronically adiabatic approximation, we see that $|\dot{\xi}\rangle \cong 0$, and likewise for $|\dot{\xi}'\rangle$, so

$$|\dot{\Psi}_R\rangle \cong |g\rangle|\dot{\psi}_g\rangle + |\xi\rangle|\dot{\psi}_\xi\rangle + |\xi'\rangle|\dot{\psi}_{\xi'}\rangle\,. \tag{4.29}$$

The instantaneous evolution of $|\Psi_R\rangle$ in the electronically adiabatic basis is seen to reside in its nuclear wave packets.

[4] What matters in signal formation is not the value of θ during the launch pulse, which, in particular, is zero when $t_d > \tau_1$, but its value during and after the Raman probe pulse (for $|\uparrow_0\uparrow_1\rangle$) and its value during the Raman probe (for $|\uparrow_0\uparrow_1\downarrow_2\rangle$).

How adiabaticity works. Consider a two-state system with the Hamiltonian $h(t) = \beta(\sigma_z \cos\phi(t) + \sigma_x \sin\phi(t)) = \beta u(t)\sigma_z u^\dagger(t)$, where $u(t) = e^{-\frac{i}{2}\phi(t)\sigma_y}$, with

$$\sigma_x = |f\rangle\langle e| + |e\rangle\langle f|\,,$$

$$\sigma_y = -i|f\rangle\langle e| + i|e\rangle\langle f|\,,$$

and

$$\sigma_z = |f\rangle\langle f| - |e\rangle\langle e|\,.$$

Let $|\bar{\psi}(t)\rangle = u^\dagger(t)|\psi(t)\rangle$, and show that $|\dot{\bar{\psi}}\rangle = \left(\frac{i}{2}\dot{\phi}\,\sigma_y - \frac{i}{\hbar}\beta\sigma_z\right)|\bar{\psi}\rangle$ follows from $|\dot{\psi}\rangle = -\frac{i}{\hbar}h(t)|\psi\rangle$. Find the evolution equation for $|\bar{\bar{\psi}}\rangle = e^{\frac{i}{\hbar}\beta t\sigma_z}|\bar{\psi}\rangle$. Suppose the angle ϕ changes at a constant rate $r \ll \beta/\hbar$, and solve for $|\psi(t)\rangle$ through first order in r using time-dependent perturbation theory. Show that $|\psi(t)\rangle \cong e^{-i\frac{rt}{2}\sigma_y}e^{-i\frac{\beta t}{\hbar}\sigma_z}|\psi(0)\rangle$ to a good approximation in this situation, even if $\phi = rt$ becomes large.

In order to calculate $|\uparrow_0\uparrow_1\rangle$ and $|\uparrow_0\uparrow_1\downarrow_2\rangle$ using time-dependent perturbation theory, both of first order in V_0 and of zeroth and first order in V_2, respectively (see above, following eqns (4.26) and (4.27)), we introduce an interaction picture, $|\tilde{\Psi}\rangle = U(t)|\Psi_R\rangle$, where $U(t) = |g\rangle e^{i\frac{t}{\hbar}H_g}\langle g| + |\xi\rangle e^{i\frac{t}{\hbar}H_e}\langle\xi| + |\xi'\rangle e^{i\frac{t}{\hbar}(H_f - \hbar\Omega_1)}\langle\xi'|$ and the initial condition is $|\tilde{\Psi}(-\tau_1)\rangle = U(-\tau_1)|\Psi_R(-\tau_1)\rangle = |g\rangle|0_g(Q)\rangle|0_{0_g}\rangle$. Using

$$\dot{U}(t) \cong \frac{i}{\hbar}\left[|g\rangle\langle g|e^{i\frac{t}{\hbar}H_g}H_g + |\xi\rangle\langle\xi|e^{i\frac{t}{\hbar}H_e}H_e + |\xi'\rangle\langle\xi'|e^{i\frac{t}{\hbar}(H_f-\hbar\Omega_1)}(H_f - \hbar\Omega_1)\right], \quad (4.30)$$

we obtain $|\dot{\tilde{\Psi}}\rangle = \{\dot{U} - \frac{i}{\hbar}U[H_R(f_1) + V_0 + V_2]\}|\Psi_R\rangle = -\frac{i}{\hbar}(\tilde{V}_0 + \tilde{V}_2)|\tilde{\Psi}\rangle$, in which

$$\tilde{V}_0(t) \cong -\frac{\mu_{eg}E_0}{2}f_0(t)\,e^{-i\varphi_0}|\xi\rangle\langle g|e^{\frac{it}{\hbar}h_e}e^{-\frac{it}{\hbar}(h_g+\hbar\Omega_0)} + H.c.\,, \quad (4.31)$$

and

$$\begin{aligned}\tilde{V}_2(t) \cong &-\frac{\mu_{fe}E_2}{2}f_2(t-t_d)\,e^{i\varphi_2 - i\varphi_1}e^{i(\Omega_2-\Omega_1)(t-t_d)}\\ &\times\left[|\xi'\rangle\langle\xi'|\,e^{\frac{i}{\hbar}t_dH_f}e^{\frac{i}{\hbar}(t-t_d)h_f}\frac{\theta}{2}\,e^{-\frac{i}{\hbar}(t-t_d)h_f}e^{-\frac{i}{\hbar}t_dH_f}\right.\\ &\left.-|\xi\rangle\langle\xi|\,e^{\frac{i}{\hbar}t_dH_e}e^{\frac{i}{\hbar}(t-t_d)h_e}\frac{\theta}{2}\,e^{-\frac{i}{\hbar}(t-t_d)h_e}e^{-\frac{i}{\hbar}t_dH_e}\right] + H.c.\end{aligned} \quad (4.32)$$

Since the launch and Raman probe pulses are abrupt on the conformational timescale, we have been able to make semi-classical Franck–Condon approximations in eqns (4.31) and (4.32) by neglecting noncommutativity between conformational kinetic and potential energy operators over the length of the pulses. In $\tilde{V}_0$, for example,

$$e^{\frac{it}{\hbar}H_e}e^{-\frac{it}{\hbar}H_g} \cong e^{\frac{it}{\hbar}h_e(Q)}e^{\frac{it}{\hbar}\frac{P^2}{2M}}e^{-\frac{it}{\hbar}\frac{P^2}{2M}}e^{-\frac{it}{\hbar}h_g(Q)} = e^{\frac{it}{\hbar}h_e}e^{-\frac{it}{\hbar}h_g}\,. \quad (4.33)$$

From the interaction-picture Schrödinger equation above eqn (4.31) together with the interaction-picture equivalent of eqn (4.29), we have $|\dot{\tilde{\psi}}_\xi\rangle_0 = -\frac{i}{\hbar}\langle\xi|\tilde{V}_0|g\rangle|\tilde{\psi}_\xi\rangle$, whence

$$
\begin{aligned}
|\tilde{\psi}_\xi(t>\tau_0)\rangle_0 &= -\frac{i}{\hbar}\int_{-\tau_0}^{\tau_0} d\tau\, \langle\xi(f_1(\tau-t_d))|\tilde{V}_0(\tau)|g\rangle|0_g(Q)\rangle|0_{0_g}\rangle \\
&= iF_0 e^{-i\varphi_0} p_0^{(eg)}|0_g(Q)\rangle|0_{0_g}\rangle\,, \qquad (4.34)
\end{aligned}
$$

where

$$
p_0^{(eg)} = \int_{-\tau_0}^{\tau_0}\frac{d\tau}{\tau_0}\, f_0(\tau)\, e^{\frac{i\tau}{\hbar}h_e} e^{-\frac{i\tau}{\hbar}(h_g+\hbar\Omega_0)}\,, \qquad (4.35)
$$

and $F_0 = \mu_{eg}E_0\tau_0/2\hbar$. In the (still rotating) Schrödinger picture,

$$
\begin{aligned}
|\Psi_R(t>\tau_0)\rangle_0 &= U^\dagger(t)|\xi(f_1(t-t_d))\rangle|\tilde{\psi}_\xi(t)\rangle_0 \\
&= |\xi(f_1(t-t_d))\rangle\, iF_0 e^{-i\varphi_0 - i\frac{t}{\hbar}H_e} p_0^{(eg)}|0_g(Q)\rangle|0_{0_g}\rangle\,, \qquad (4.36)
\end{aligned}
$$

from which we obtain

$$
\begin{aligned}
|\Psi(t>\tau_0)\rangle_0 &= R^\dagger(t)|\Psi_R(t)\rangle_0 \\
&= \left[|e\rangle - |f\rangle\frac{1}{2}\,\theta(f_1(t-t_d))\, e^{-i\varphi_1 - i(t-t_d)\Omega_1}\right] \\
&\quad \times iF_0 e^{-i\varphi_0 - i\frac{t}{\hbar}H_e} p_0^{(eg)}|0_g(Q)\rangle|0_{0_g}\rangle \\
&= |\uparrow_0\rangle + |\uparrow_0\uparrow_1\rangle\,. \qquad (4.37)
\end{aligned}
$$

Hence, we recognize

$$
|\uparrow_0\uparrow_1\rangle = -|f\rangle\, iF_0\, e^{-i\varphi_0 - i\varphi_1 - i(t-t_d)\Omega_1}\frac{\theta(f_1(t-t_d))}{2}\, e^{-i\frac{t}{\hbar}H_e} p_0^{(eg)}|0_g(Q)\rangle|0_{0_g}\rangle\,, \qquad (4.38)
$$

as the f-state amplitude created by the launch pulse and the subsequent action of the Raman pump. Because the Raman pump field is subresonant with $f \leftarrow e$, this amplitude relies for its existence on the continuing presence of the pump envelope f_1.

We shall assume that $\hbar\Omega_0$ is chosen to match the $0_e \leftarrow 0_g$ vibronic transition energy (viz., $\hbar\Omega_0 \cong v_e(0) - v_g(0)$). Hence, in eqn (4.38),

$$
\begin{aligned}
e^{-i\frac{t}{\hbar}H_e} p_0^{(eg)}|0_g(Q)\rangle &\cong e^{-i\frac{t}{\hbar}H_e}|0_e(Q)\rangle\langle 0_e(Q)|p_0^{(eg)}|0_g(Q)\rangle \\
&= |0_e(Q)\rangle e^{-\frac{it}{\hbar}\left(\frac{P^2}{2M}+\hbar\epsilon_{0_e}(Q)\right)}\langle 0_e(Q)|p_0^{(eg)}|0_g(Q)\rangle\,. \qquad (4.39)
\end{aligned}
$$

As a result of this simplification, eqn (4.38) becomes

$$
\begin{aligned}
|\uparrow_0\uparrow_1\rangle &= -|f\rangle\, iF_0\, e^{-i\varphi_0 - i\varphi_1 - i(t-t_d)\Omega_1}\frac{\theta(f_1(t-t_d))}{2}|0_e(Q)\rangle \\
&\quad \times \exp\left\{-\frac{it}{\hbar}\left(\frac{P^2}{2M} + \hbar\epsilon_{0_e}(Q)\right)\right\}\langle 0_e(Q)|p_0^{(eg)}|0_g(Q)\rangle|0_{0_g}\rangle\,, \qquad (4.40)
\end{aligned}
$$

in which we see that the evolution of this two-pulse ket is determined by the dynamics of the conformational wave packet in the $\hbar\epsilon_{0_e}(Q)$ potential.

In search of the e-state amplitude generated by the Raman probe pulse, we integrate $|\dot{\tilde{\psi}}_\xi\rangle_{02} = -\frac{i}{\hbar}\langle\xi|\tilde{V}_2|\xi\rangle|\tilde{\psi}_\xi\rangle_0$ to obtain

$$|\tilde{\psi}_\xi(t > t_d - \tau_2)\rangle_{02} = -i\frac{\mu_{fe}E_2}{2\hbar}e^{i\varphi_2 - i\varphi_1 + \frac{i}{\hbar}t_d H_e} \tag{4.41}$$
$$\times \int_{t_d-\tau_2}^{t} d\tau\, f_2(\tau - t_d)\, e^{\frac{i}{\hbar}(\tau - t_d)(h_e + \hbar\Omega_2)} \frac{\theta(f_1(\tau - t_d))}{2} e^{-\frac{i}{\hbar}(\tau - t_d)(h_e+\hbar\Omega_1)}\, e^{-\frac{i}{\hbar}t_d H_e}|\tilde{\psi}_\xi(\tau)\rangle_0$$
$$= F_0 F_2 e^{i\varphi_2 - i\varphi_1 - i\varphi_0} e^{\frac{i}{\hbar}t_d H_e}\zeta(t - t_d)e^{-\frac{i}{\hbar}t_d H_e} p_0^{(eg)}|0_g(Q)\rangle|0_{0_g}\rangle\,,$$

where

$$\zeta(t - t_d) = \frac{1}{2}\int_{-\tau_2}^{t-t_d}\frac{d\tau}{\tau_2}\, f_2(\tau)\, e^{i\tau(\Omega_2 - \Omega_1)} e^{\frac{i\tau}{\hbar}h_e(Q)}\, \theta(1)\, e^{-\frac{i\tau}{\hbar}h_e(Q)}\,, \tag{4.42}$$

and $F_2 = \mu_{fe}E_2\tau_2/2\hbar$. In defining ζ we have taken advantage of the fact that $f_1(\tau) \cong f_1(0) = 1$ for τ between $-\tau_2$ and τ_2. In the Schrödinger picture, for $t > t_d - \tau_2$,

$$|\Psi_R(t)\rangle_{02} = |\xi(f_1(t - t_d))\rangle\langle\xi(f_1(t - t_d))|U^\dagger(t)|\xi(f_1(t - t_d))\rangle|\tilde{\psi}_\xi(t)\rangle_{02} \tag{4.43}$$
$$= |\xi(f_1(t - t_d))\rangle F_0 F_2 e^{i\varphi_2 - i\varphi_1 - i\varphi_0 - \frac{i}{\hbar}(t - t_d)H_e}\zeta(t - t_d)e^{-\frac{i}{\hbar}t_d H_e}p_0^{(eg)}|0_g(Q)\rangle|0_{0_g}\rangle\,.$$

Finally, in the nonrotating frame we find

$$|\uparrow_0\uparrow_1\downarrow_2\rangle = |e\rangle\langle e|R^\dagger(t)|\Psi_R(t)\rangle_{02} \tag{4.44}$$
$$= |e\rangle F_0 F_2 e^{i\varphi_2 - i\varphi_1 - i\varphi_0 - \frac{i}{\hbar}(t - t_d)H_e}\zeta(t - t_d)e^{-\frac{i}{\hbar}t_d H_e}p_0^{(eg)}|0_g(Q)\rangle|0_{0_g}\rangle$$
$$\cong |e\rangle F_0 F_2 e^{i\varphi_2 - i\varphi_1 - i\varphi_0 - \frac{i}{\hbar}(t - t_d)H_e}\zeta(t - t_d)|0_e(Q)\rangle$$
$$\times \exp\Big\{-\frac{it_d}{\hbar}\Big(\frac{P^2}{2M} + \hbar\epsilon_{0_e}(Q)\Big)\Big\}\langle 0_e(Q)|p_0^{(eg)}|0_g(Q)\rangle|0_{0_g}\rangle\,.$$

We shall see that, unlike $|\uparrow_0\uparrow_1\rangle$, this three-pulse e-state amplitude can outlive the envelope of the Raman pump pulse.

Let's develop an approximate expression for $\theta/2$ that will help in isolating the part of the fissors dipole responsible for Raman gain on the $1_e \leftarrow 0_e$ vibrational transition. By expanding this operator, defined in eqn (4.24), in the vibrational potential-energy shift between the e- and f-states, we get

$$\frac{\theta(f_1)}{2} = -\frac{\mu_{fe}E_1/2}{H_f - \hbar\Omega_1 - H_e}f_1$$
$$\cong -\mathcal{F}_1 f_1\Big\{1 + \frac{m\omega^2(Q)}{2\hbar(\epsilon_{fe} - \Omega_1)}\big[2\delta q(q - q_e(Q)) - \delta q^2\big]\Big\}, \tag{4.45}$$

where $\mathcal{F}_1 \equiv \frac{\mu_{fe}E_1}{2\hbar(\epsilon_{fe} - \Omega_1)}$.

Work out this expansion.

Next, we can employ vibrational creation and annihilation operators to write $q - q_e(Q) = \sqrt{\frac{\hbar}{2m\omega(Q)}}\big(a_e^\dagger(Q) + a_e(Q)\big)$ and from eqn (4.42) thereby obtain

$$\zeta(t - t_d) \cong -\mathcal{F}_1 \Delta(Q) \int_{-\tau_2}^{t-t_d} \frac{d\tau}{\tau_2}\, f_2(\tau)\, e^{i\tau(\Omega_2+\omega(Q)-\Omega_1)}\, a_e^\dagger(Q)\,, \tag{4.46}$$

with $\Delta(Q) \equiv \frac{m\omega^2(Q)}{2\hbar(\epsilon_{fe}-\Omega_1)} 2\delta q \sqrt{\frac{\hbar}{2m\omega(Q)}}$. In this formula, use has been made of $\tau_2 > 2\pi/(\Omega_1 - \Omega_2) \approx 2\pi/\omega(Q)$ to neglect a nonresonant term involving $a_e(Q)$. Since the launch pulse is incapable of preparing vibrational superpositions in the e-state ($\tau_0 > 2\pi/\omega(0)$) and, by our choice of Ω_0 it prepares amplitude in $|0_e\rangle$, the presence of $a_e^\dagger$ in eqn (4.46) implies that $|\uparrow_0\uparrow_1\downarrow_2\rangle$ has significant amplitude only in $|1_e(Q)\rangle$.

In calculating μ_{FSRS}, it is therefore sufficient to replace eqn (4.40) with

$$\begin{aligned} |1_e(Q)\rangle\langle 1_e(Q)|\uparrow_0\uparrow_1\rangle = &-|f\rangle|1_e(Q)\rangle\, iF_0\, e^{-i\varphi_0-i\varphi_1-i(t-t_d)\Omega_1} \\ &\times \langle 1_e(Q)|\frac{\theta(f_1(t-t_d))}{2}\, e^{-i\frac{t}{\hbar}H_e} p_0^{(eg)}|0_g(Q)\rangle|0_{0_g}\rangle \\ \cong &\ |f\rangle|1_e(Q)\rangle\, iF_0\mathcal{F}_1 e^{-i\varphi_0-i\varphi_1-i(t-t_d)\Omega_1} f_1(t-t_d)\Delta(Q) \\ &\times \exp\Big\{-\frac{it}{\hbar}\Big(\frac{P^2}{2M} + \hbar\epsilon_{0_e}(Q)\Big)\Big\}\langle 0_e(Q)|p_0^{(eg)}|0_g(Q)\rangle|0_{0_g}\rangle\,. \end{aligned} \tag{4.47}$$

A similar refinement can be made in the three-pulse amplitude of eqn (4.44) as well, giving

$$\begin{aligned} &|1_e(Q)\rangle\langle 1_e(Q)|\uparrow_0\uparrow_1\downarrow_2\rangle = |e\rangle|1_e(Q)\rangle F_0F_2 e^{i\varphi_2-i\varphi_1-i\varphi_0}\, e^{-\frac{i}{\hbar}(t-t_d)\left(\frac{P^2}{2M}+\hbar\epsilon_{1_e}(Q)\right)} \\ &\times \langle 1_e(Q)|\zeta(t-t_d)|0_e(Q)\rangle e^{-\frac{i}{\hbar}t_d\left(\frac{P^2}{2M}+\hbar\epsilon_{0_e}(Q)\right)}\langle 0_e(Q)|p_0^{(eg)}|0_g(Q)\rangle|0_{0_g}\rangle \\ &\cong -|e\rangle|1_e(Q)\rangle F_0\mathcal{F}_1F_2 e^{i\varphi_2-i\varphi_1-i\varphi_0} \exp\Big\{-\frac{i}{\hbar}(t-t_d)\Big(\frac{P^2}{2M} + \hbar\epsilon_{1_e}(Q)\Big)\Big\} \\ &\quad\times \Delta(Q)\int_{-\tau_2}^{t-t_d} \frac{d\tau}{\tau_2}\, f_2(\tau)\, e^{i\tau(\Omega_2+\omega(Q)-\Omega_1)} \\ &\quad\times \exp\Big\{-\frac{i}{\hbar}t_d\Big(\frac{P^2}{2M} + \hbar\epsilon_{0_e}(Q)\Big)\Big\}\langle 0_e(Q)|p_0^{(eg)}|0_g(Q)\rangle|0_{0_g}\rangle\,. \end{aligned} \tag{4.48}$$

Equations (4.47) and (4.48) express the two-pulse f-state wave packet and three-pulse e-state wave packet, respectively, and hence the fissors dipole (eqn (4.10)), in terms of *the dynamics of the conformational degree of freedom alone.* This feature makes manifest the role of FSRS as a probe of conformational motion.

We denote the conformational wave packet at the arrival time of the Raman pump and probe as

$$|\chi(t_d)\rangle = \exp\Big\{-\frac{i}{\hbar}t_d\Big(\frac{P^2}{2M} + \hbar\epsilon_{0_e}(Q)\Big)\Big\}\langle 0_e(Q)|p_0^{(eg)}|0_g(Q)\rangle|0_{0_g}\rangle\,. \tag{4.49}$$

This packet has propagated on $\hbar\epsilon_{0_e}(Q)$ since the launch pulse. By combining eqns (4.47) and (4.48) in eqn (4.10), we obtain a final working expression for the fissors dipole:

$$\mu_{FSRS}(t,t_d) = 2\mu_{fe}F_0^2\mathcal{F}_1^2F_2f_1(t-t_d)\,\mathrm{Re}\Big[i\,e^{i\varphi_2+i(t-t_d)\Omega_1} \tag{4.50}$$
$$\times \langle\chi(t_d)|\exp\Big\{\frac{i}{\hbar}(t-t_d)\Big(\frac{P^2}{2M}+\hbar\epsilon_{0_e}(Q)\Big)\Big\}\,\Delta(Q)$$
$$\times \exp\Big\{-\frac{i}{\hbar}(t-t_d)\Big(\frac{P^2}{2M}+\hbar\epsilon_{1_e}(Q)\Big)\Big\}\,\Delta(Q)\int_{-\tau_2}^{t-t_d}\frac{d\tau}{\tau_2}\,f_2(\tau)\,e^{i\tau(\Omega_2+\omega(Q)-\Omega_1)}|\chi(t_d)\rangle\Big]\,.$$

Except for the weakly Q-dependent factors $\Delta(Q)$ and the integral involving the envelope of the Raman-probe pulse, which reduces to its Fourier transform, $\frac{1}{\tau_2}\tilde{f}_2(\Omega_2+\omega(Q)-\Omega_1)$, when $t-t_d$ exceeds τ_2, eqn (4.50) gives μ_{FSRS} as the inner product between two copies of the conformational wave packet $|\chi(t_d)\rangle$. One of these has evolved for $t-t_d$ on $\hbar\epsilon_{1_e}(Q)$ and the other has evolved for this time on $\hbar\epsilon_{0_e}(Q)$. The fissors dipole does not "turn on" until $t-t_d=-\tau_2$, because f_2 vanishes beforehand.[5]

In carrying out FSRS signal calculations, as we shall do in the next section, it is helpful to specify the form of the pulse envelopes appearing in eqn (4.8). We adopt the simple truncated-cosine functions

$$f_I(t) = \begin{cases} \cos\left(\frac{\pi t}{2\tau_I}\right) & \text{for } -\tau_I < t < \tau_I \\ 0 & \text{otherwise} \end{cases}\,. \tag{4.51}$$

1. Using the envelope function of eqn (4.51), find a closed-form expression for $\langle 0_e(Q)|p_0^{(eg)}|0_g(Q)\rangle$, which appears in eqns (4.47) and (4.48) and obtain, in particular, an explicit formula for the Franck–Condon overlap $\langle 0_e(Q)|0_g(Q)\rangle$. Describe the qualitative conformational wave-packet-shaping effect of the launch pulse in light of the brevity of τ_0 relative to $2\pi/\Omega$ and the forms of $v_g(Q)$ and $v_e(Q)$ (see below). Note that it is only through this reduced pulse propagator that the Q-dependent displacement $q_e(Q)$ contributes to the fissors dipole.
2. Work out a formula for the integral $\int_{-\tau_2}^{t-t_d}\frac{d\tau}{\tau_2}\,f_2(\tau)\,e^{i\tau\Omega}$ appearing in eqn (4.48). Take account of the fact that $t-t_d$ may either be between $-\tau_2$ and τ_2 or greater than τ_2.

4.4 Example signal calculation

4.4.1 Numerical treatment of dynamical model

For the purpose of illustration, we adopt a model Hamiltonian that could represent a diatomic molecule adsorbed to a solid surface. In eqns (4.1) through (4.3), we choose the conformational potentials,

$$v_g(Q) = \frac{M\Omega^2}{2}Q^2\,, \tag{4.52}$$

[5] The fact that eqn (4.50) for the fissors dipole is indifferent to the form of the Raman-pump envelope prior to $t=-\tau_2$, provided it changes electronically adiabatically, is consistent with the assertion made in the Fig. 4.1 caption that our treatment applies regardless of whether or not the launch pulse is embedded within the pump.

$$v_e(Q) = \frac{M\Omega^2}{2} Q^2 \,\Theta(-Q) + \hbar\epsilon_{eg} \,, \tag{4.53}$$

$$v_f(Q) = \frac{M\Omega^2}{2} Q^2 \,\Theta(-Q) + \hbar\epsilon_{fg} \,. \tag{4.54}$$

The step function $\Theta(-Q)$ zeroes out the restoring force for positive Q, allowing the molecule to desorb from the surface once it is electronically excited.[6] The dependence of the vibrational frequency and equilibrium displacements on Q are specified by $\omega(Q) = \omega(1 - \frac{1}{3}\tanh(\frac{Q}{4Q_{rms}}))$, $q_g(Q) = 0$, $q_e(Q) = \frac{1}{2} q_{rms} \tanh(\frac{Q}{4Q_{rms}})$, and $q_f(Q) = q_e(Q) + \frac{1}{4} q_{rms}$, where $Q_{rms} = \sqrt{\frac{\hbar}{2M\Omega}}$ and $q_{rms} = \sqrt{\frac{\hbar}{2m\omega}}$; the molecule's vibrational frequency is lower and its equilibrium bond length longer when it is further from the surface. We take $M = 40m$ and—in support of vibrational adiabaticity—let $\Omega = \omega/12$. We set $\epsilon_{eg} = 45\omega$ and $\epsilon_{fg} = 80\omega$, whence $\epsilon_{fe} = 35\omega$.

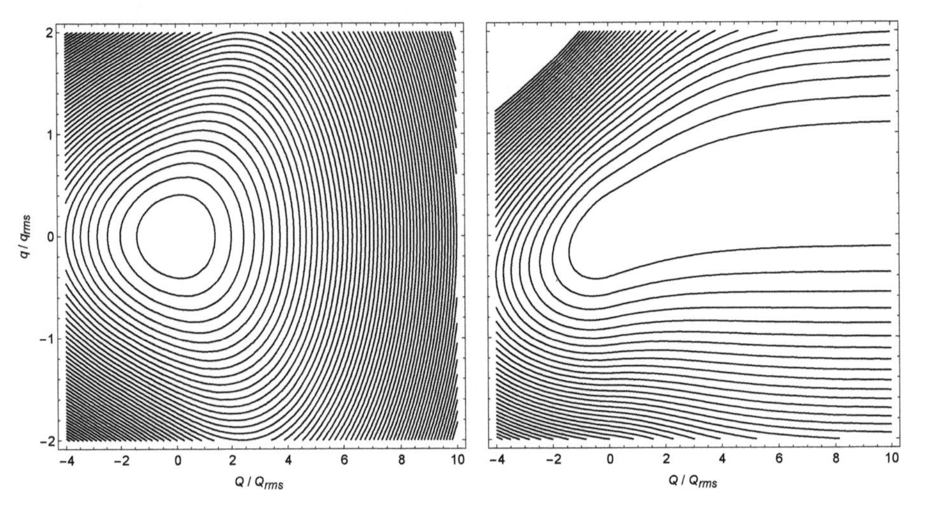

Fig. 4.2 Electronic potential energy surfaces governing nuclear motion in the g-state (left) and e-state (right). Contours are spaced $\hbar\Omega/2$. Visible in the right-hand panel are the increased equilibrium displacement of the vibrational mode and its decreased vibrational frequency for positive values of the conformational coordinate. The f-state potential surface has the same form as the e-state, but is shifted upward by $0.25q_{rms}$.

The two-dimensional electronic potential energy surfaces $\frac{m\omega^2(Q)}{2}\left(q - q_j(Q)\right)^2 + v_j(Q)$ for $j = g, e$ appearing in the nuclear Hamiltonians of eqn (4.3) are shown in Fig. 4.2. The f-state potential coincides with that for the e-state, shifted upward by $\delta q = q_{rms}/4$. Under the vibrationally adiabatic approximation, conformational dynamics is governed, not by the $v_j(Q)$ alone, but by the potential energy functions

[6] Note that if the conformational motion were *classical*, transfer from v_g to v_e at $Q = 0$ would not induce desorption: it is the nonzero width of the conformational wave packet in the electronic ground state that allows it to respond to the repulsive potential in the e-state.

$\hbar\epsilon_{n_j}(Q) = \hbar\omega(Q)(n_j + \frac{1}{2}) + v_j(Q)$ of eqn (4.5). These potential curves for the relevant 0_g, 0_e, and 1_e vibronic states are plotted in Fig. 4.3.

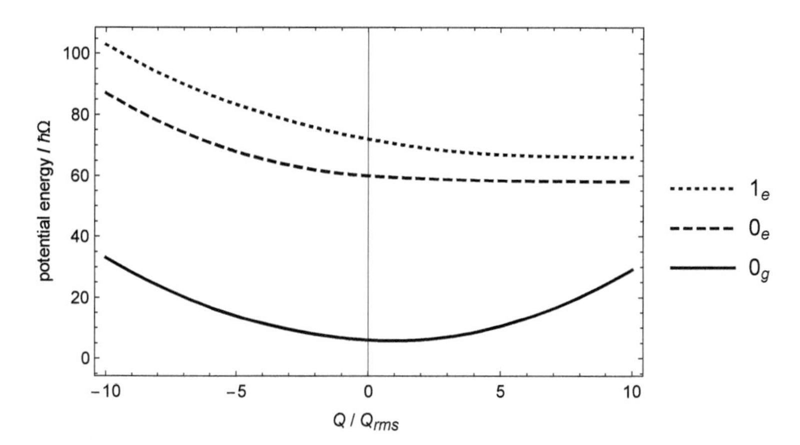

Fig. 4.3 Conformational potentials $\hbar\epsilon_{0_g}(Q)$, $\hbar\epsilon_{0_e}(Q)$, and $\hbar\epsilon_{1_e}(Q)$, with the latter two shifted downward in energy to fit on the same plot. Note that the minimum position in $\hbar\epsilon_{0_g}(Q)$ is shifted from $Q = 0$ to a slightly larger value by the conformational-coordinate dependence of the vibrational zero-point energy.

Find the equilibrium value of the conformational coordinate in $\hbar\epsilon_{0_g}(Q)$. What is the vibrational vibrational frequency $\omega(Q)$ at this conformation? What is $q_e(Q)$?

The center frequency of the launch pulse, $\Omega_0 = \epsilon_{eg}$, is resonant with the $e \leftarrow g$ transition, and its duration parameter, $\tau_0 = 1.2(2\pi/\omega)$, is just a bit longer then the vibrational period. The Raman pump pulse, with $\Omega_1 = \epsilon_{fe} - 7\omega$, is below resonance with $f \leftarrow e$ and is assigned a duration $\tau_1 = 2.0(2\pi/\Omega)$ in the time-range of *conformational* change. The Raman probe has $\Omega_2 = \Omega_1 - \omega$, and its pulse-length, $\tau_2 = 1.4(2\pi/\omega)$, slightly in excess of the vibrational period, ensures that the pump and probe together selectively excite the $1_e \leftarrow 0_e$ vibrational transition by a stimulated Raman process. The parameter determining the spectral resolution of the detector, introduced below eqn (4.12), is set at $\delta\omega = \omega/40$. We do not specify values for the transition dipole moments or the electric field strengths and simply state the fissors signal, which is proportional to $(\mu_{eg}E_0)^2(\mu_{fe}E_1)^2(\mu_{fe}E_2)^2$, in arbitrary units.

The conformational motion corresponding to the molecule's ejection from the surface to which it was adsorbed, following electronic excitation by the launch pulse, is shown in Fig. 4.4. The probability density $|\langle Q|\chi(t_d)\rangle|^2$ is plotted after evolving for $t_d = 0.0$, 0.8, and $1.6 \times (2\pi/\Omega)$ in the desorptive $\hbar\epsilon_{0_e}(Q)$ potential. This is the otherwise difficult-to-access dynamical process one hopes to observe through FSRS.

The fissors signal of this photo-desorptive model system as a function of t_d and $\bar{\omega}$ can be calculated from eqn (4.16). The requisite dipole function is given by eqn (4.10) as an overlap between the two-pulse ket of eqn (4.47) and the three-pulse ket of eqn (4.48). It can be evaluated as a function of t_d and t by numerically evaluating the

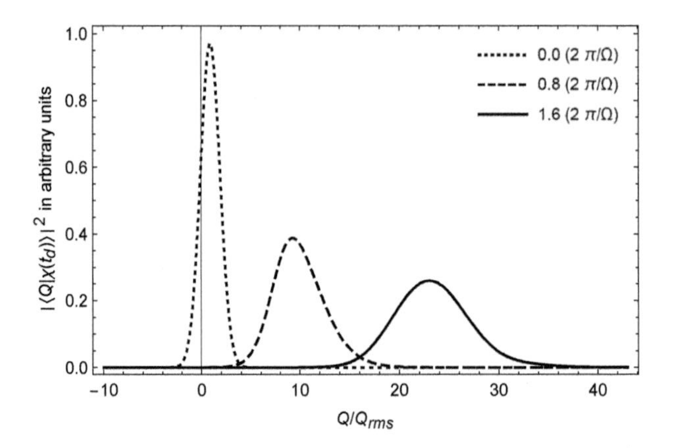

Fig. 4.4 Conformational probability density for various t_d of propagation in $\hbar\epsilon_{0_e}(Q)$.

one-dimensional conformational dynamics under $\frac{P^2}{2M} + \hbar\epsilon_{0_g}(Q)$, $\frac{P^2}{2M} + \hbar\epsilon_{0_e}(Q)$, and $\frac{P^2}{2M} + \hbar\epsilon_{1_e}(Q)$. This task was accomplished by finding eigenvalues and eigenfunctions of these Hamiltonians (as well as the other Q-dependent factors in the fissors dipole) in a common discrete conformational coordinate basis.[7] In the final integration of eqn (4.16), the time-increment can under-sample the vibronic oscillations, as those of the fissors dipole and the probe field cancel, resulting in an integrand that varies meaningfully only on the conformational timescale.

The calculated time-resolved fissors spectrum shown in Fig. 4.5 clearly manifests the underlying conformational dynamics. At the shortest delay times, the peak Raman shift is about $0.93\,\omega$, consistent with the average conformational coordinate $Q \approx 0.95\,Q_{rms}$ in 0_g.[8] The broad initial $0.3\,\omega$ range of Raman shifts corresponds to the approximately $3\,Q_{rms}$-wide distribution of the conformational coordinate in the ground-state of the adsorbed molecule. This distribution narrows markedly and red-shifts to a peak Raman shift of $0.667\,\omega$ by $t_d \approx 2\pi/\Omega$, matching the vibrational frequency of the desorbed species.

The sliver of negative signal for $t_d < 0.5\,(2\pi/\Omega)$, appearing in purple in Fig. 4.5, results from a conformationally induced change in the "instantaneous frequency" of the fissors dipole that is rapid relative to that dipole's disappearance. This frequency shift leads to destructive interference between the red-edge Fourier components of the field radiated by the fissors dipole and the Raman-probe field, which gives rise to

[7]This basis spanned a spatial range from $Q = -10\,Q_{rms}$ to $Q = 250\,Q_{rms}$ with a grid spacing of $Q_{rms}/12$. In this approach, each conformational wave packet is expressed as a sum over component eigenfunctions weighted by t-dependent phase factors. The replacement of a continuous e-state conformational energy eigenspectrum with a discrete one is inconsequential for the time-ranges of interest. To speed the calculations, the e-state eigenbases were truncated at about six hundred states.

[8]Note, however, that due to its neglect of launch-probe overlap, our treatment is not strictly valid for $t_d < \tau_0 + \tau_2 \approx 0.22\,(2\pi/\Omega)$.

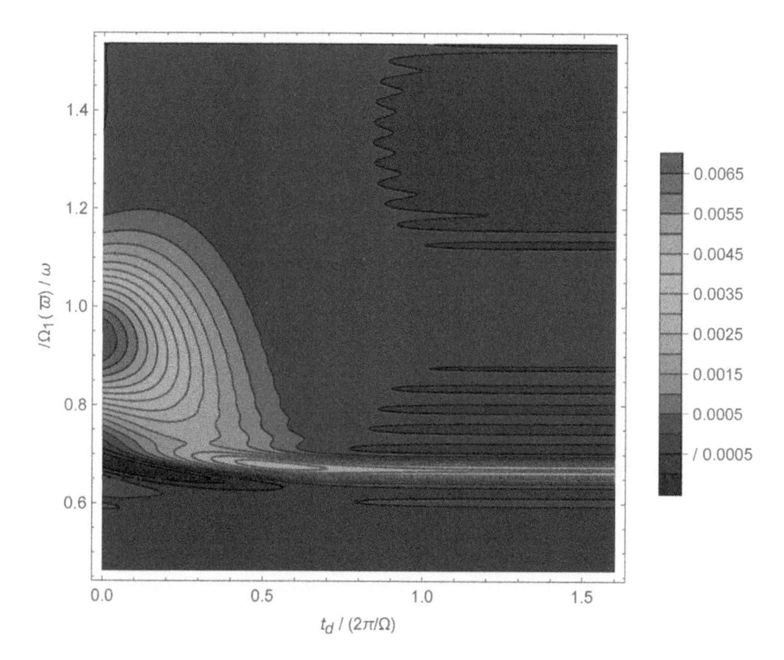

Fig. 4.5 The fissors spectrum as a contour plot of the observed signal in arbitrary units with respect to the post-launch delay t_d and the Raman shift $\Omega_1 - \bar{\omega}$. The spectrum is plotted over the bandwidth of the Raman probe pulse, with $\bar{\omega}$ spanning the range $\Omega_2 \pm 3\pi/2\tau_2$.

negative Raman gain (see the following subsection).[9]

The fissors spectrum does not necessarily provide a "snapshot" of the conformational motion. Although the Raman pump and probe envelopes peak simultaneously at t_d, and the $t = 0$ value of $\mu_{FSRS}(t, t_d)$ is set by $\langle Q|\chi(t_d)\rangle$, that conformational wave packet continues to evolve, and does so differently on $\hbar\epsilon_{0_e}(Q)$ and $\hbar\epsilon_{1_e}(Q)$. The signal-generating fissors dipole persists for $t > t_d$ until it is attenuated by vibrational decoherence (i.e., loss of overlap between the 0_e- and 1_e-associated packets) or extinguished by the completion of the pump-pulse envelope.

Both sources of dipole decay are seen Fig. 4.6, which plots the magnitude of the fissors dipole for the case $t_d = 0.25(2\pi/\Omega)$ over the time-range $t_d - \tau_2 < t < t_d + \tau_1$. It begins to appear with the advent of the Raman probe, at $t = t_d - \tau_2$, but starts to diminish even before the end of the pulse at $t = t_d + \tau_2 = 0.367(2\pi/\Omega)$, due to the differing acceleration of the conformational wave packets in the Franck–Condon region of $\hbar\epsilon_{0_e}(Q)$ and $\hbar\epsilon_{1_e}(Q)$ (consult Fig. 4.3). This diminution, which constitutes vibrational decoherence, is abated after $t \approx 0.5(2\pi/\Omega)$, when both wave packets are in the desorbed region. The envelope of the Raman-pump pulse governs the subsequent, gradual decay of the fissors dipole in this model system.[10]

[9] I am grateful to Professor David W. McCamant, University of Rochester, for his comments on early drafts of this chapter and specifically for his help in apprehending the significance of this interesting feature.

[10] In many physical systems, the fissors dipole disappears before the end of the Raman pump as a result of stronger vibrational dephasing in the presence of multiple low-frequency bath modes.

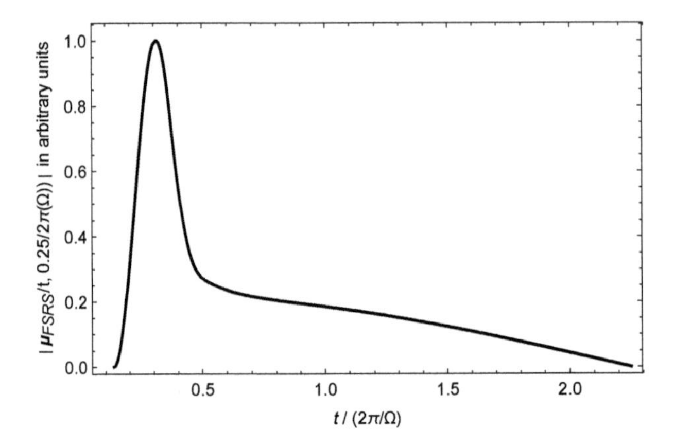

Fig. 4.6 Absolute value of μ_{FSRS} as a function of t for $t_d = 0.25(2\pi/\Omega)$.

Can you speculate about the form and relative size of the fissors dipole in a case, such as $t_d = 1.0(2\pi/\Omega)$, where the conformational wave packets have left their respective Franck–Condon regions before the beginning of the Raman-probe pulse?

The evolving size of μ_{FSRS} tells only part of the story of the measured Raman gain. The delay-dependent shift in the peak of the fissors spectrum in Fig. 4.5, from $\Omega_1 - \bar{\omega} \cong 0.93\,\omega$ to $0.67\,\omega$, evinces a change in the temporal *phase structure* of the fissors dipole, born of the "instantaneous" vibrational frequency's dependence on the changing conformation.

4.4.2 Simplified description

In order to illustrate some salient consequences of the fissors dipole's phase development, we entertain a simplified, analytical form for $\mu_{FSRS}(t, t_d)$.[11] If we neglect the Q-dependence of Δ in eqn (4.50), ignore the non-zero duration of the Raman probe, and replace the truncated-cosine pump-pulse envelope with the exponential form $f_1(t) = e^{-|t|/\tau_1}$, we may then write[12]

$$\mu_{FSRS}(t, t_d) \cong \mu_{t_d}\Theta(t - t_d)\,\mathrm{Re}\big\{i\, e^{i\varphi_2 + i(t-t_d)(\Omega_1 + i/\tau_1) - i\varphi(t) + i\varphi(t_d)}\big\}\,. \tag{4.55}$$

In this formula, μ_{t_d} is a real-valued prefactor, and $\Theta(t)$ is the unit step-function. We further assume that the derivative of the phase function tracks an exponentially shifting time-dependent frequency, so that

$$\frac{d}{dt}\varphi(t) = \omega(t) = \omega_0 + \Delta\omega(1 - e^{-t/\tau_c})\,, \tag{4.56}$$

[11] FSRS signals based on a dipolar source with an exponentially shifting "instantaneous frequency," similar to those considered here, are studied in: D. W. McCamant, "Re-evaluation of rhodopsin's relaxation kinetics determined from femtosecond stimulated Raman lineshapes," J. Phys. Chem. B **115**, 9299–9305 (2011) and S. Mukamel and J. D. Biggs, "Comment on the effective temporal and spectral resolution of impulsive stimulated Raman signals," J. Chem. Phys. **134**, 161101/1–4 (2011).

[12] Alternatively, τ_1 can be regarded as a vibrational decoherence time.

where the "conformational relaxation time" τ_c sets the timescale over which the oscillation frequency shifts from ω_0 to $\omega_0+\Delta\omega$ due, as here for instance, to the conformation-dependence of a vibrational frequency. Integrating eqn (4.56) gives

$$\varphi(t) - \varphi(t_d) = (\omega_0 + \Delta\omega)(t - t_d) + \Delta\omega\tau_c\, e^{-t_d/\tau_c}(e^{-(t-t_d)/\tau_c} - 1)\,. \tag{4.57}$$

We may now investigate a fissors-like signal by working from eqn (4.16). Making the additional assumption of very high spectral resolution ($\delta\omega \ll 2\pi/\tau_1$), we arrive at a signal expression of the form

$$S(\bar{\omega}) = \frac{\mu_{t_d}}{2}\,\mathrm{Re}\int_0^\infty d\tau\ \exp\left\{i\tau(\Omega + i/\tau_1) - is(t_d)(e^{-\tau/\tau_c} - 1)\right\}, \tag{4.58}$$

where $\Omega \equiv \Omega_1 - \bar{\omega} - \omega_0 - \Delta\omega$ and $s(t_d) \equiv \Delta\omega\,\tau_c\, e^{-t_d/\tau_c}$.

In the case of rapid conformational change, $s(t_d)$ is small for all delays because $|\Delta\omega\,\tau_c| \ll 1$, and eqn (4.58) can be approximated as

$$S(\bar{\omega}) \cong \frac{\mu_{t_d}}{2}\,\mathrm{Re}\int_0^\infty d\tau\ \left\{[1 + is(t_d)]e^{i\tau(\Omega+i/\tau_1)} - is(t_d)e^{i\tau(\Omega+i/\tau_1+i/\tau_c)}\right\}. \tag{4.59}$$

Since $1/\tau_c > \Delta\omega > 1/\tau_1$ in the present instance, the second term in braces does not contribute significantly, and upon integration eqn (4.59) gives

$$S(\bar{\omega}) = \frac{\mu_{t_d}}{2}\,\frac{(1/\tau_1) - \Omega\, s(t_d)}{\Omega^2 + (1/\tau_1)^2}\,. \tag{4.60}$$

Near the unshifted vibrational resonance, we can write $\bar{\omega} = \Omega_1 - \omega_0 + \delta\bar{\omega}$ with some small increment $\delta\bar{\omega}$, and

$$S(\Omega_1 - \omega_0 + \delta\bar{\omega}) \cong \frac{\mu_{t_d}}{2}\,\frac{(1/\tau_1) + (\Delta\omega + \delta\bar{\omega})s(t_d)}{(\Delta\omega + \delta\bar{\omega})^2}\,. \tag{4.61}$$

Regardless of which term in the numerator of eqn (4.61) predominates, the lineshape at the unshifted resonance is not strongly asymmetric. In the vicinity of the fully shifted vibrational resonance, we have $\bar{\omega} = \Omega_1 - \omega_0 - \Delta\omega + \delta\bar{\omega}$, and eqn (4.60) becomes

$$S(\Omega_1 - \omega_0 - \Delta\omega + \delta\bar{\omega}) \cong \frac{\mu_{t_d}}{2}\,\frac{(1/\tau_1) + \delta\bar{\omega}\, s(t_d)}{\delta\bar{\omega}^2 + (1/\tau_1)^2}\,. \tag{4.62}$$

At long delays, $s(t_d)$ goes to zero, and the signal,

$$S \cong \frac{\mu_{t_d}}{2}\,\frac{1/\tau_1}{\delta\bar{\omega}^2 + (1/\tau_1)^2}\,, \tag{4.63}$$

becomes a Lorentzian line centered at the fully shifted resonance, just as seen in Fig. 4.5. At short delay-times, though, provided τ_c is not *so* short as to prevent $|\delta\bar{\omega}\,\tau_c\,\Delta\omega| > 1/\tau_1$, we see different behavior, as

$$S \cong \frac{\mu_{t_d}}{2}\,\frac{\tau_c\Delta\omega e^{-t_d/\tau_c}}{\delta\bar{\omega}}\,. \tag{4.64}$$

In this situation, the short-delay FSRS spectrum near the fully shifted resonance is negative (positive) when $\delta\bar{\omega}$ and $\Delta\omega$ differ (are the same) in sign. This result explains the negative feature at t_d less than τ_c seen in Fig. 4.5.

Only in the case of extremely sluggish conformational motion, where τ_c exceeds τ_1, does the fissors signal directly monitor the changing vibrational frequency. Here, we can make the approximation $e^{-\tau/\tau_c} - 1 \cong -\frac{\tau}{\tau_c} - \frac{\tau^2}{2\tau_c^2}$ and complete the square in the exponent of eqn (4.58) to obtain

$$S(\bar{\omega}) \cong \frac{\mu_{t_d}}{2} \mathrm{Re}\left\{ e^{-i\frac{s(t_d)}{2}\left[1+\frac{\tau_c}{s(t_d)}\left(\Omega+\frac{i}{\tau_1}\right)\right]^2} \int_{\tau_c\left[1+\frac{\tau_c}{s(t_d)}\left(\Omega+\frac{i}{\tau_1}\right)\right]}^{\infty} d\tau \, e^{i\frac{s(t_d)}{2\tau_c^2}\tau^2} \right\}. \qquad (4.65)$$

Since the argument of the exponent in the integrand is large throughout the range of integration, we can write

$$e^{is\tau^2/2\tau_c^2} = -i\frac{\tau_c^2}{s}\left[\frac{d}{d\tau}\left(\frac{1}{\tau}e^{is\tau^2/2\tau_c^2}\right) + \frac{1}{\tau^2}e^{is\tau^2/2\tau_c^2}\right], \qquad (4.66)$$

and neglect the much smaller, second term inside the square brackets. Integration then leads to

$$S(\bar{\omega}) \cong \frac{\mu_{t_d}}{2} \frac{1/\tau_1}{\left(\Omega_1 - \bar{\omega} - \omega(t_d)\right)^2 + 1/\tau_1^2}, \qquad (4.67)$$

a Lorentzian-shaped resonance centered at $\Omega_1 - \bar{\omega} = \omega(t_d)$ (see eqn (4.56)). In this slow-motion limit of ponderous conformational change, the fissors spectrum tracks the delay-dependent vibrational frequency.

Under the condition $\tau_c \approx 1/\Delta\omega$, where conformational motion is neither relatively rapid nor very slow, we can resort to rigorous evaluation of the integral in eqn (4.58) by writing

$$\begin{aligned} S(\bar{\omega}) &= \frac{\mu_{t_d}}{2} \sum_{n=0}^{\infty} \frac{1}{n!} \mathrm{Re}\left\{ e^{is(t_d)}\left(-is(t_d)\right)^n \int_0^{\infty} d\tau \, e^{i\tau\left(\Omega + \frac{i}{\tau_1} + \frac{in}{\tau_c}\right)} \right\} \\ &= \frac{\mu_{t_d}}{2} \sum_{n=0}^{\infty} \frac{1}{n!} \mathrm{Re}\left\{ e^{is(t_d)}\left(-is(t_d)\right)^n \frac{1}{\Omega + \frac{i}{\tau_1} + \frac{in}{\tau_c}} \right\}; \end{aligned} \qquad (4.68)$$

the summand becomes negligible for n somewhat larger than $\Delta\omega\,\tau_c$.

Write a program to evaluate the idealized delay-dependent FSRS spectrum using eqn (4.68), and explore its form over a representative range of parameter values.

The model molecular system examined in this section shares with the one considered in the preceding chapter on transient-absorption spectroscopy the presence of one higher- and one lower-frequency nuclear degree of freedom. In the interest of generality, the transient-absorption chapter did not take explicit advantage of this time-scale disparity to develop a vibrationally adiabatic approximation in which the eigenstates of the high frequency mode are parametrized by the coordinate operator of the low frequency mode. The next chapter revisits transient-absorption spectroscopy in this more specialized situation. The resulting treatment affords an opportunity to illustrate the added information content of fissors relative to the more widely practiced method of transient absorption.

5

Transient-absorption reprise: Taking advantage of vibrational adiabaticity

Chapter 3 on transient-absorption spectroscopy presents signal calculations for a two-mode system consisting of one high- and one low-frequency nuclear degree of freedom but does not avail itself of this timescale separation to develop a vibrationally adiabatic treatment. The approach taken in that chapter would apply in the more general situation where the target chromophore contains or couples to multiple nuclear modes with a range of frequencies but exhibits no clear break between slow modes and fast ones. In this chapter, we build on the material from Chapter 4 to develop a theoretical description of transient-absorption measurements on molecules of the kind studied there, which again feature a high-frequency "vibration" and a low-frequency "conformational coordinate." This time through, we make use of vibrational adiabaticity and also compare the physical information content of calculated fissors and transient-absorption signal plots for the same system.

5.1 Transient-absorption signal under vibrational adiabaticity

5.1.1 Signal expression

The model Hamiltonian, $H = |g\rangle H_g\langle g| + |e\rangle H_e\langle e| + |f\rangle H_f\langle f|$, is the same as given in eqn (4.1), and the nuclear Hamiltonians, with eigenstates and eigenenergies incorporating vibrational adiabaticity, are those detailed in eqns (4.2) through (4.6). We shall investigate the transient-absorption signal using the time-dependent Hamiltonian, $H(t) = H - \hat{\mu}E(t)$, with an electronic dipole operator specified by eqn (4.7), and the time-dependent field,

$$E(t) = E_1(t) + E_2(t)\,, \tag{5.1}$$

comprising the laser electric fields of the pump (1) and probe (2) taking the form given by eqn (4.8). The arrival time of the pump pulse is $t_1 = 0$, and $t_2 = t_d$ is the pump-probe delay. The center frequency of the pump pulse, with $\hbar\Omega_1 \approx \upsilon_e - \upsilon_g$, is now resonant with $e \leftarrow g$. In a "one-color" case, the probe frequency Ω_2 is similar to Ω_1. In a "two-color" case with $\hbar\Omega_2 \approx \upsilon_f - \upsilon_e$, it is resonant with the lower-frequency $f \leftrightarrow e$ transition. Transient-absorption spectroscopy prizes time resolution almost capable of "freezing" vibrational motion and benefitting from the concomitant spectral breadth,[1] so, in contrast to fissors, we here assume τ_1, $\tau_2 < 2\pi/\omega$ (see eqn (4.51) and page 47).

[1]For example, see C. C. Jumper, I. H. M. van Stokkum, T. Mirkovic, and G. D. Scholes, "Vibronic wavepackets and energy transfer in cryptophyte light-harvesting complexes," J. Phys. Chem. B **122**, 6328–6340 (2018), and D. Polli, P. Altoè, O. Weingart, K. M. Spillane, C. Manzoni, D. Brida, G.

By analogy with the fissors signal expression of eqn (4.16) or from the earlier direct derivation of eqn (3.16), we can write the transient-absorption signal as

$$\Delta\mathcal{U}(\bar{\omega}) = -E_2'\Omega_2 \int_{-\infty}^{\infty} d\tau\, f_2'(\tau - t_d) \sin[\bar{\omega}(\tau - t_d) + \varphi_2]\, \mu_{TA}(\tau)\,; \tag{5.2}$$

E_2' and f_2' are the field amplitude and envelope function, respectively, of the spectrally filtered transmitted probe pulse defined below eqn (4.13). The transient-absorption dipole, proportional to $E_1^2 E_2$, can be transcribed from eqn (3.18) in the form

$$\mu_{TA}(t) = 2\mathrm{Re}\{\underbrace{\langle\Psi_0|\hat{\mu}|\uparrow_1\downarrow_1\uparrow_2\rangle + \langle\Psi_0|\hat{\mu}|\uparrow_2\downarrow_1\uparrow_1\rangle}_{\text{GSB}} + \underbrace{\langle\uparrow_1\downarrow_2|\hat{\mu}|\uparrow_1\rangle}_{\text{SE}} + \underbrace{\langle\uparrow_1\downarrow_1|\hat{\mu}|\uparrow_2\rangle}_{\text{IRE}}\}\,, \tag{5.3}$$

for the one-color case, and

$$\mu_{TA}(t) = 2\mathrm{Re}\{\underbrace{\langle\uparrow_1|\hat{\mu}|\uparrow_1\uparrow_2\rangle}_{\text{ESA}}\}\,, \tag{5.4}$$

for the two-color case, where $|\Psi_0\rangle$ is the unperturbed molecular state at time t.

What would become of eqns (5.3) and (5.4) in a situation where Ω_1 is *sub*-resonant with $e \leftarrow g$ and therefore unable to permanently transfer amplitude from the ground to the first excited electronic state?

5.1.2 Bras and kets

Formulas for the multi-pulse kets appearing in eqns (5.3) and (5.4) are readily obtained using time-dependent perturbation theory, along lines detailed in Chapters 3 and 4. For one-color measurements, we need

$$|\uparrow_1\rangle = |e\rangle\, iF_1^{(eg)} e^{-i\varphi_1} [t]_{ee}\, p_1^{(eg)}(t;\tau)|0_g\rangle\,, \tag{5.5}$$

where $F_1^{(eg)} = E_1\mu_{eg}\tau_1/2\hbar$, and $|0_g\rangle$, obeying $H_g|0_g\rangle = \hbar\mathcal{E}_{0_g}|0_g\rangle$, stands for some initial nuclear eigenket in the electronic ground state;

$$|\uparrow_2\rangle = |e\rangle\, iF_2^{(eg)} e^{-i\varphi_2} [t - t_d]_{ee}\, p_2^{(eg)}(t - t_d;\tau)[t_d]_{gg}|0_g\rangle\,, \tag{5.6}$$

with $F_2^{(eg)} = E_2\mu_{eg}\tau_2/2\hbar$;

$$|\uparrow_1\downarrow_1\rangle = -|g\rangle\, (F_1^{(eg)})^2 [t]_{gg}\, p_1^{(ge)}(t;\tau) p_1^{(eg)}(\tau;\bar{\tau})|0_g\rangle\,; \tag{5.7}$$

$$|\uparrow_1\downarrow_2\rangle = -|g\rangle\, F_1^{(eg)} F_2^{(eg)} e^{i\varphi_2 - i\varphi_1} [t - t_d]_{gg}\, p_2^{(ge)}(t - t_d;\tau)[t_d]_{ee}\, p_1^{(eg)}(\tau + t_d;\bar{\tau})|0_g\rangle\,; \tag{5.8}$$

$$\begin{aligned} |\uparrow_1\downarrow_1\uparrow_2\rangle = &-|e\rangle\, i(F_1^{(eg)})^2 F_2^{(eg)} e^{-i\varphi_2} \\ &\times [t - t_d]_{ee}\, p_2^{(eg)}(t - t_d;\tau)[t_d]_{gg}\, p_1^{(ge)}(\tau + t_d;\bar{\tau}) p_1^{(eg)}(\bar{\tau};\bar{\bar{\tau}})|0_g\rangle\,; \end{aligned} \tag{5.9}$$

Tomasello, G. Orlandi, P. Kukura, R. A. Mathies, M. Garavelli, and G. Cerullo, "Conical intersection dynamics of the primary photoisomerization event in vision," Nature **467**, 440–443 (2010).

and

$$|\uparrow_2\downarrow_1\uparrow_1\rangle = -|e\rangle\, i(F_1^{(eg)})^2 F_2^{(eg)} e^{-i\varphi_2}$$
$$\times\, [t]_{ee}\, p_1^{(eg)}(t;\tau) p_1^{(eg)}(\tau;\bar{\tau})[-t_d]_{ee}\, p_2^{(eg)}(\bar{\tau}-t_d;\bar{\bar{\tau}})[t_d]_{gg}|0_g\rangle\,. \tag{5.10}$$

In the two-color case, we need $|\uparrow_1\rangle$, just as in eqn (5.5), and

$$|\uparrow_1\uparrow_2\rangle = -|f\rangle\, F_1^{(eg)} F_2^{(fe)} e^{-i\varphi_2 - i\varphi_1}$$
$$\times\, [t-t_d]_{ff}\, p_2^{(fe)}(t-t_d;\tau)[t_d]_{ee}\, p_1^{(eg)}(\tau+t_d;\bar{\tau})|0_g\rangle\,, \tag{5.11}$$

with $F_2^{(fe)} = E_2\mu_{fe}\tau_2/2\hbar$.

5.2 Calculated transient-absorption signals

5.2.1 Set-up under vibrationally adiabatic approximation

In this section, we present the set-up for calculations of both one- and two-color transient-absorption signals for the photo-desorptive molecular system whose fissors spectrum was investigated in Section 4.4.1.[2] We then present the results of such calculations for the one-color case alone. A convenient basis for that system is $\{|j\rangle|n_j(Q)\rangle|m_{n_j}\rangle\}$, which respects its vibrational adiabaticity. The arbitrary initial nuclear state referred to above as $|0_g\rangle$ will be taken to be the lowest-lying nuclear eigenstate in the electronic ground state, denoted as $|0_g(Q)\rangle|0_{0_g}\rangle$, with eigenenergy $\hbar\mathcal{E}_{0_{0_g}}$. The nuclear free-evolution operators,

$$[t]_{jj} = \langle j|[t]|j\rangle \cong \sum_{n_j} |n_j(Q)\rangle e^{-i\frac{t}{\hbar}\{\frac{P^2}{2M}+\hbar\epsilon_{n_j}(Q)\}}\langle n_j(Q)|\,, \tag{5.12}$$

are taken to be vibrationally diagonal on the assumption that conformational change is too slow to effect nonadiabatic vibrational transitions.

Because both of the laser pulses are brief on the vibrational timescale ($\tau_I < 2\pi/\omega$), it is certainly safe to simplify the reduced pulse propagators with a semi-classical Franck–Condon approximation that neglects non-commutativity of the *conformational* potential and kinetic energy operators. Hence, for example,[3]

$$\langle n_e(Q)|\, p_I^{(eg)}(t;\tau)|n_g(Q)\rangle \cong \langle n_e(Q)|n_g(Q)\rangle \tag{5.13}$$
$$\times \int_{-\infty}^{t} \frac{d\tau}{\tau_I}\, f_I(\tau)\exp\big\{-i\tau\big(\Omega_I + \epsilon_{n_g}(Q) - \epsilon_{n_e}(Q)\big)\big\}\,.$$

For the simple choice of pulse-envelope function given in eqn (4.51) and adopted here, the time integral in this reduced pulse propagator can be evaluated as

[2] In this chapter, we double the vibrational Franck–Condon displacement between the e and f electronic states to $q_f(Q) - q_e(Q) = \delta q = q_{rms}/2$, in order to strengthen slightly the vibronic progression associated with excited-state absorption.

[3] Compare this simplified formula with the more general eqn (3.21).

$$\int_{-\infty}^{t} \frac{d\tau}{\tau_I} f_I(\tau)\, e^{-i\tau\Omega} = \frac{\Theta(t-\tau_I)}{(\pi/2)^2-(\tau_I\Omega)^2}\, \pi \cos\tau_I\Omega \tag{5.14}$$
$$+ \frac{\Theta(t+\tau_I)\Theta(\tau_I-t)}{(\pi/2)^2-(\tau_I\Omega)^2}\left\{ e^{-it\Omega}\left[\frac{\pi}{2}\sin\Big(\frac{\pi t}{2\tau_I}\Big) - i\tau_I\Omega\cos\Big(\frac{\pi t}{2\tau_I}\Big)\right] + \frac{\pi}{2}\, e^{i\tau_I\Omega}\right\}.$$

A formula for $\langle n_f(Q)|\, p_I^{(fe)}(t;\tau)|n_e(Q)\rangle$ similar to eqn (5.13) follows immediately.

Derive eqns (5.13) and (5.14). Does the latter feel familiar?

At the cost of making errors at delay-times shorter than the pulse durations, which in the present situation are *very* brief on the timescale of conformational dynamics, we neglect temporal overlap between the pump and probe pulses. In the one-color case, this neglect entails making the replacement

$$p_2^{(ge)}(t-t_d;\tau)[t_d]_{ee}\, p_1^{(eg)}(\tau+t_d;\bar{\tau}) \cong p_2^{(ge)}(t-t_d;\tau)[t_d]_{ee}\, p_1^{(eg)}(\infty;\bar{\tau})\,, \tag{5.15}$$

in the stimulated-emission overlap of eqn (5.3), along with

$$p_2^{(eg)}(t-t_d;\tau)[t_d]_{gg}\, p_1^{(ge)}(\tau+t_d;\bar{\tau})p_1^{(eg)}(\bar{\tau};\bar{\bar{\tau}}) \tag{5.16}$$
$$\cong p_2^{(eg)}(t-t_d;\tau)[t_d]_{gg}\, p_1^{(ge)}(\infty;\bar{\tau})p_1^{(eg)}(\bar{\tau};\bar{\bar{\tau}})\,,$$

in the ground-state-bleach term, and neglecting $\langle\Psi_0|\hat{\mu}|\uparrow_2\downarrow_1\uparrow_1\rangle$ entirely. In the two-color case, we make the replacement

$$p_2^{(fe)}(t-t_d;\tau)[t_d]_{ee}\, p_1^{(eg)}(\tau+t_d;\bar{\tau}) \cong p_2^{(fe)}(t-t_d;\tau)[t_d]_{ee} p_1^{(eg)}(\infty;\bar{\tau})\,, \tag{5.17}$$

in the single, excited-state-absorption element $\langle\uparrow_1|\hat{\mu}|\uparrow_1\uparrow_2\rangle$ of eqn (5.4).

We are left with one doubly nested reduced pulse propagator, which appears in the impulsive-Raman-excitation element of eqn (5.3) for the one-color signal. By inserting a completeness relation using adiabatic vibrational eigenstates in the first electronic excited state and making conformational semiclassical Franck–Condon approximations in both inner and outer integrals, we obtain

$$\langle n_g(Q)|\, p_1^{(ge)}(\infty;\tau)\, p_1^{(eg)}(\tau;\bar{\tau})|0_g(Q)\rangle \cong \frac{1}{2}\sum_{n_e=0}^{\infty}\langle n_g(Q)|n_e(Q)\rangle\langle n_e(Q)|0_g(Q)\rangle$$
$$\times\left\{ i\sin\tau_1(\Omega-\bar{\Omega})\left[\tfrac{1}{\tau_1\bar{\Omega}-\frac{\pi}{2}}\left(\tfrac{1}{\tau_1(\Omega-\bar{\Omega})}-\tfrac{1}{\tau_1(\Omega-\bar{\Omega})+\pi}\right)+\tfrac{1}{\tau_1\bar{\Omega}+\frac{\pi}{2}}\left(\tfrac{1}{\tau_1(\Omega-\bar{\Omega})}-\tfrac{1}{\tau_1(\Omega-\bar{\Omega})-\pi}\right)\right]\right.$$
$$\left. +\,\pi^2 e^{i\tau_1\bar{\Omega}}\cos\tau_1\Omega\, \tfrac{1}{\left[(\frac{\pi}{2})^2-\tau_1^2\Omega^2\right]^2\left[(\frac{\pi}{2})^2-\tau_1^2\bar{\Omega}^2\right]^2}\right\}, \tag{5.18}$$

where $\Omega = \Omega_1 + \epsilon_{n_g}(Q) - \epsilon_{n_e}(Q)$ and $\bar{\Omega} = \Omega_1 + \epsilon_{0_g}(Q) - \epsilon_{n_e}(Q)$; note that $\Omega - \bar{\Omega} = \epsilon_{n_g}(Q) - \epsilon_{0_g}(Q)$ is the overall vibrational transition frequency.

Verify eqn (5.18).

In carrying out numerical signal calculations on this system, it is helpful (e.g., in generating the Franck–Condon overlaps appearing in matrix elements of the reduced pulse propagators) to express adiabatic e-state vibrational eigenkets in terms of the g-state eigenkets and the f-state vibrational eigenkets in terms of the e-state kets. These tasks are readily accomplished through power-series expansions of vibrational spatial-translation operators. For example, since $|n_e(Q)\rangle = \exp\{-iq_e(Q)p/\hbar\}|n_g(Q)\rangle$, where in this instance n_e and n_g are understood to be the same integer, we can write $p = i\sqrt{m\hbar\omega(Q)/2}\,(a_g^\dagger(Q) - a_g(Q))$, expand the exponential, and use operator algebra to obtain

$$|n_e(Q)\rangle = \sqrt{n_e!}\, e^{-\frac{m\omega(Q)}{4\hbar}q_e^2(Q)} \sum_{m_g=0}^{\infty} |m_g(Q)\rangle \sqrt{m_g!} \tag{5.19}$$
$$\times \sum_{l=0}^{\min(n_e,m_g)} \frac{(-1)^{n_e-l}}{(n_e-l)!\,l!\,(m_g-l)!} \left[q_e(Q)\sqrt{\frac{m\omega(Q)}{2\hbar}}\right]^{n_e+m_g-2l}.$$

Since the displacement in the equilibrium vibrational coordinate between the f- and e-states is the constant value $q_f(Q) - q_e(Q) = \delta q$, we have similarly,

$$|n_f(Q)\rangle = \sqrt{n_f!}\, e^{-\frac{m\omega(Q)}{4\hbar}\delta q^2} \sum_{m_e=0}^{\infty} |m_e(Q)\rangle \sqrt{m_e!} \tag{5.20}$$
$$\times \sum_{l=0}^{\min(n_f,m_e)} \frac{(-1)^{n_f-l}}{(n_f-l)!\,l!\,(m_e-l)!} \left[\delta q\sqrt{\frac{m\omega(Q)}{2\hbar}}\right]^{n_f+m_e-2l}.$$

Derive eqn (5.19). Can you produce the converse formula, for $|n_g(Q)\rangle$ in terms of the $\{|m_e(Q)\rangle\}$, without rederivation?

5.2.2 Results

Here, we calculate only two-color transient-absorption signals for this model system, with the probe pulse resonantly accessing the $f \leftarrow e$ electronic transition, as these can be compared most directly with the fissors spectrum of Fig. 4.5. Figure 5.1 shows signals with $\Omega_1 = \epsilon_{eg}$, $\Omega_2 = \epsilon_{fe}$, and $\tau_1 = \tau_2 = 0.2(2\pi/\omega)$ (i.e., a fifth of the vibrational period) for interpulse delays $t_d = 0.0(2\pi/\Omega)$ and $1.0(2\pi/\Omega)$ (i.e., zero and one times the conformational period in the electronic ground state). The spectral-resolution parameter is set to $\delta\omega = 0.5(2\pi/\Omega)$ and the plotted range of the observed frequency is $\Omega_2 - \pi/2\tau_2 < \bar{\omega} < \Omega_2 + \pi/2\tau_2$, which spans the middle third of the probe-pulse power spectrum.[4,5]

[4]Recall that we are calculating the pump-induced change in the electromagnetic energy of the Fourier amplitude of the transmitted probe pulse within a variable narrow spectral range. Despite referring to this quantity as the "transient absorption," greater absorption of probe-pulse energy by the molecule leads to a more *negative* value of the signal.

[5]Conformational eigenstates and eigenenergies in each vibronic level were found by diagonalizing the Hamiltonians $P^2/2M + \hbar\epsilon_{n_j}(Q)$ on a one-dimensional position grid from $-36\,Q_{rms}$ to $254\,Q_{rms}$

At both delays, the signal consists of a large central peak corresponding to excited-state absorption at frequencies $\bar{\omega} \approx \Omega_2 = \epsilon_{fe}$, with additional weak but discernible structure at $\bar{\omega} \approx \epsilon_{fe} + 0.53(\pi/2\tau_2) \cong \epsilon_{fe} + 0.66\omega$ (along with barely visible wiggles at $\bar{\omega} \approx \epsilon_{fe} - 0.53(\pi/2\tau_2) \cong \epsilon_{fe} - 0.66\omega$). In order to apprehend this seemingly sparse result, we first examine the pump-driven $e \leftarrow g$ transition. The pump pulse has a bandwidth, $\sim 2\pi/\tau_1 = 5\omega$, more than sufficient to generate vibrational superpositions in the e-state. But eqn (5.19) states that $\langle 0_g(Q)|0_e(Q)\rangle = e^{-m\omega(Q)q_e^2(Q)/4\hbar}$ and $\langle 0_g(Q)|1_e(Q)\rangle = \langle 0_g(Q)|0_e(Q)\rangle\, q_e(Q)\sqrt{m\omega(Q)/2\hbar}$; the latter Franck–Condon overlap vanishes at $Q = 0$ and remains very small for $|Q| < Q_{rms}$. Hence, despite its brevity and concomitant broad spectral width, the pump pulse transfers the initial vibrational state $|0_g(Q)\rangle$ predominantly to $|0_e(Q)\rangle$. Because the $f \leftarrow e$ Franck–Condon overlap $\langle 0_e(Q)|1_f(Q)\rangle = \langle 0_e(Q)|0_f(Q)\rangle\, \delta q\sqrt{m\omega(Q)/2\hbar}$ involves the constant displacement $\delta q = 0.5q_{rms}$, the 0-0 bias is much less pronounced for the probe-driven excited-state absorption transitions. The strong $\bar{\omega} \approx \epsilon_{fe}$ peaks in both $t_d = 0.0(2\pi/\Omega)$ and $1.0(2\pi/\Omega)$ spectra result mostly from 0_e-to-0_f transitions, while the small blips at $\bar{\omega} - \Omega_2 \cong 0.66\omega$ can be assigned to the 0_e-to-1_f vibronic transition of the molecule in its desorbed conformation (and those at $\bar{\omega} - \Omega_2 \cong -0.66\omega$ belong to the very weak 1_e-to-0_f transition).

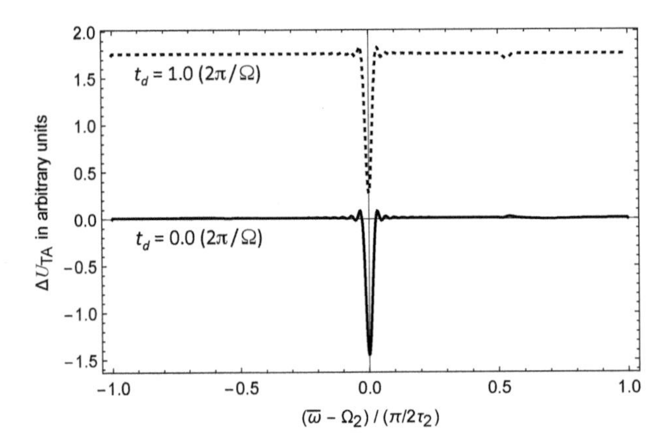

Fig. 5.1 Two-color transient-absorption spectra for the photo-desorption model at pump-probe delays of zero and one times the ground-state conformational period. The two traces are vertically displaced for clarity.

Direct information on conformational dynamics comes to light when we zero in on the detailed structure of the small 0_e-to-1_f peak. This can be done by plotting the part of each spectrum that is antisymmetric under a change in sign of $\bar{\omega} - \Omega_2$, and these are shown in Fig. 5.2. Whereas the $t_d = 2\pi/\Omega$ spectrum displays decreased probe transmission at the precise frequency of the 0_e-to-1_f vibronic transition for the desorbed conformation, excited-state vibronic absorption in the shorter-delay trace occurs over

with a spacing of $\Delta Q = Q_{rms}/12$. The resulting conformational eigenbases were then truncated at 600 states each.

a range of $\bar{\omega} - \Omega_2$ from $0.6\pi/2\tau_2$ to $0.85\pi/2\tau_2$ (0.75 to 1.06 times ω) corresponding to the vibrational frequencies $\omega(Q)$ encompassed by the g-state conformational wave function. This behavior is a consequence of the e-state conformational dynamics illustrated in Fig. 4.4 and the Q-dependence of the 0_e-to-1_f transition energy. For, from eqn (4.5) we have $\epsilon_{1_f}(Q) - \epsilon_{0_e}(Q) = \epsilon_{fe} + \omega(Q)$; at the shortest pump-probe delays this transition frequency reflects the conformational-coordinate distribution of the photo-excited species still maintaining its adsorbed conformation, while at longer delays it occurs at the vibronic transition frequency of the desorbed molecule. We see that a treatment based on vibrational adiabaticity helps us interpret the subtle influences of conformational dynamics on the two-color transient-absorption signal that are more vividly manifested in the fissors spectrum of Fig. 4.5.

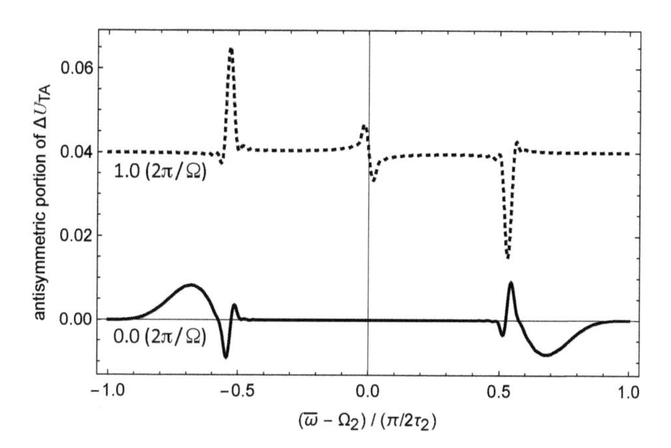

Fig. 5.2 The antisymmetric component of the spectra displayed in Fig. 5.1, accentuating the contribution of the 0_e-to-1_f transition to excited-state absorption.

The vibrational brevity of the pump pulse gives rise to a dynamical feature that is absent in the fissors spectrum. There is a *small* amount of vibrational coherence in the form of a superposition between $|0_e(Q)\rangle$ and a tiny amplitude in $|1_e(Q)\rangle$ within the pump-excited vibrational wave packet (see above). As a result of this coherence, the two-dimensional vibrational-conformational wave packet in the e-state slaloms slightly as the molecule desorbs, executing small-amplitude vibrational oscillations, rather than simply sliding down the bottom of the trough shown in the right panel of Fig. 4.2.[6] The $\bar{\omega} \approx \Omega_2$ excited-state-absorption band therefore peaks alternately slightly to the red and slightly to the blue of ϵ_{fe} with increasing pump-probe delay. This behavior is demonstrated in Fig. 5.3, which plots the $t_d = \frac{2\pi}{\Omega}$ signal shown in Fig. 5.2 along with that for a delay-time incremented by half a vibrational period of the desorbed molecule ($t_d = \frac{2\pi}{\Omega} + 0.5\,\frac{2\pi}{2\omega/3} = 1.0625\,\frac{2\pi}{\Omega}$). The half-period delay increment leads to a shift from a small blue-side excess of $\bar{\omega} \approx \Omega_2$ excited-state absorption to a small red-side excess. This feature is a two-color manifestation of a "phase-flip" in vibrational

[6] Analogous behavior is seen in the high-frequency mode of the model system considered in Chap. 3, as shown in the lower panel of Fig. 3.5.

sidebands that has been observed in one-color broadband transient-absorption data.[7]

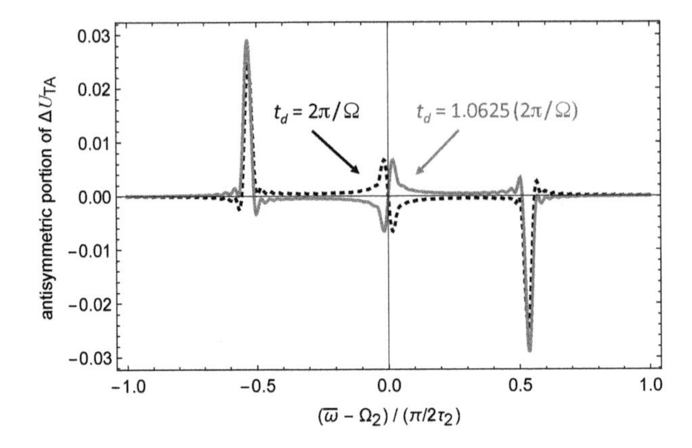

Fig. 5.3 Antisymmetric component of the transient-absorption spectrum at t_d equal to one g-state conformational period (dotted) and one conformational period plus half a vibrational period of the desorbed molecule (gray), manifesting anti-phased vibrational sidebands resulting from pump-induced vibrational coherence in the e-state.

> Consider a molecule with two electronic states, g and e, having nuclear Hamiltonians $H_j = \frac{P^2}{2M} + v_j(Q)$, where $v_g(Q) = \frac{M\Omega^2}{2}Q^2$ and $v_e(Q)$ is a particle-in-a-box potential having $v_e = \infty$ for Q less than zero or greater than L and $v_e = \hbar\epsilon_{eg}$ for $0 < Q < L$. Find the matrix elements $\langle\chi_e|p_1^{(eg)}(t,\tau)|0_g\rangle$ of the reduced pulse propagator between eigenstates of H_e and H_g, and write a program to forward-propagate the resulting e-state nuclear wave packet. How do its initial form and subsequent motion depend on the chosen carrier frequency Ω_1, pulse-duration parameter τ_1, and box-length L?

[7] J. A. Cina *et al.* (2016), cited on page 18; D. B. Turner and P. C. Arpin, "Basis set truncation further clarifies vibrational coherence spectra," Chem. Phys. **539**, 110948/1–8 (2020).

6

Two and a half approaches to two-dimensional wave-packet interferometry

6.1 Introduction

This chapter explores the description of two general strategies for *two-dimensional wave-packet interferometry* measurements having similar, but not identical, information content. These are fluorescence-detected wave-packet interferometry (which we'll sometimes denote simply as WPI, pronounced "whoopee") and heterodyne-detected four-wave mixing (FWM). Finally, we'll consider a second, popular variant of FWM based on *spectral interferometry*.[1]

Both approaches make use of four ultrashort optical pulses A, B, C, and D which are resonant with an $e \leftarrow g$ electronic transition. This sequence comprises two phase-locked pulse-pairs, AB and CD; within each pair of laser pulses there is a definite, experimentally controlled optical phase shift.

The Hamiltonian for a molecule of fixed location $\mathbf{r}$ and unchanging orientation targeted for a WPI experiment takes the form

$$H = |g\rangle H_g \langle g| + |e\rangle(\hbar\Lambda + H_e)\langle e| \,, \tag{6.1}$$

where the nuclear eigenstates and eigenenergies obey $H_j|n_j\rangle = \hbar\omega_{n_j}|n_j\rangle$ and Λ is the bare electronic transition frequency. The molecule's interaction with the pulse sequence is governed by the time-dependent Hamiltonian $H(t) = H + V(t)$ with

$$V(t) = \sum_I V_I(t) \,, \tag{6.2}$$

$$V_I(t) = -\hat{\mathbf{m}} \cdot \mathbf{E}_I(t) \,, \tag{6.3}$$

$$\hat{\mathbf{m}} = \mathbf{m}\left(|e\rangle\langle g| + |g\rangle\langle e|\right) \,, \tag{6.4}$$

and

$$\mathbf{E}_I(t) = \mathbf{e}_I E_I f_I\left(t - t_I(\mathbf{r})\right) \cos\left[\Omega(t - t_I(\mathbf{r})) + \varphi_I\right] \,. \tag{6.5}$$

The sum in eqn (6.2) is over all four pulses for a WPI experiment, but just A, B, and C for FWM. In the latter case, the D pulse plays the special role of a *local oscillator*

[1] Special thanks to my colleague Cathy Wong, who listened patiently to my ruminations on the topics of this chapter and, by declining to accept some of them at face value, helped me avoid a serious conceptual error.

(LO) and need not interact directly with the sample. In eqn (6.5) $\mathbf{e}_I$ is the polarization unit vector (assumed to be real) and $t_I(\mathbf{r}) = t_I + \mathbf{n}_I \cdot \mathbf{r}/c$ is the arrival time of the pulse at the molecular location, with t_I being the arrival time at some origin within the sample and $\mathbf{n}_I$ being a unit vector in the direction of the laser beam's spatial propagation.

We are specializing to the common situation in which the optical phase *differences* $\varphi_{BA} \equiv \varphi_B - \varphi_A$ and $\varphi_{DC} \equiv \varphi_D - \varphi_C$ are under experimental control, even though the individual φ_I may vary randomly on successive laser shots due to mechanical jitter in the optical set-up. The variable intrapulse-pair delays $t_{BA} \equiv t_B - t_A$ and $t_{DC} \equiv t_D - t_C$, or their conjugate frequency variables, are the two "dimensions" of a 2D-WPI experiment.[2,3,4,5]

6.2 Measured quantities

6.2.1 Fluorescence-detected wave-packet interferometry

The measured quantity in fluorescence-detected WPI is the quadrilinear portion of the time- and frequency-integrated fluorescence—that part proportional to $E_A E_B E_C E_D$—which in turn is presumed proportional to the quadrilinear e-state population prior to radiative or nonradiative decay. Since each pulse-induced transition changes the electronic state (from g to e or *vice versa*), e-state *amplitudes* contributing to the quadrilinear population can only result from one-pulse or three-pulse transitions (up or up-down-up, respectively).

We threrefore seek perturbative expressions for the one-pulse amplitudes

$$| \uparrow_A \rangle, \; | \uparrow_B \rangle, \; | \uparrow_C \rangle, \text{ and } | \uparrow_D \rangle \tag{6.6}$$

and the potentially contributing three-pulse amplitudes

[2]The capability to phase lock femtosecond pulse pairs was first demonstrated, and the basic concepts of wave-packet interferometry were first explored, in N. F. Scherer, R. Carlson, A. Matro, M. Du, A. J. Ruggiero, V. Romero-Rochin, J. A. Cina, G. R. Fleming, and S. A. Rice "Fluorescence-detected wave packet interferometry: Time resolved molecular spectroscopy with sequences of femtosecond phase-locked pulses," J. Chem. Phys. **95**, 1487–1511 (1991), and N. F. Scherer, A. Matro, R. J. Carlson, M. Du, L. D. Ziegler, J. A. Cina, and G. R. Fleming, "Fluorescence-detected wave packet interferometry II: Role of rotations and determination of the susceptibility," J. Chem. Phys. **96**, 4180–4194 (1992).

[3]See also A. W. Albrecht, J. D. Hybl, S. M. Gallagher Faeder, and D. M. Jonas, "Experimental distinction between phase shifts and time delays: Implications for femtosecond spectroscopy and coherent control of chemical reactions." J. Chem. Phys. **111**, 10934–10956 (1999).

[4]We do not cover the important field of multidimensional vibrational spectroscopy, which is addressed in the early works Y. Tanimura and S. Mukamel, "Two-dimensional femtosecond vibrational spectroscopy of liquids," J. Chem. Phys. **99**, 9496–9511 (1993), and P. Hamm, M. Lim, and R. M. Hochstrasser, "Structure of the amide I band of peptides measured by femtosecond nonlinear-infrared spectroscopy," J. Phys. Chem. B **102**, 6123–6138 (1998). See also P. Hamm and M. T. Zanni, *Concepts and Methods of 2D Infrared Spectroscopy* (Cambridge University Press, Cambridge, 2011).

[5]Efforts are underway to develop multi-color multidimensional spectroscopy techniques, which, for instance, simultaneously target both electronic and vibrational transitions. See J. D. Gaynor, A. Petrone, X. S. Li, and M. Khalil, "Mapping vibronic couplings in a solar cell dye with polarization selective two-dimensional electronic-vibrational spectroscopy," J. Phys. Chem. Lett. **9**, 6289–6295 (2018), and M. Cho and G. R. Fleming, "Two-dimensional electronic-vibrational spectroscopy reveals cross-correlation between solvation dynamics and vibrational spectral diffusion," J. Phys. Chem. B **124**, 11222–11235 (2020).

$$
\begin{aligned}
&|\uparrow_B\downarrow_C\uparrow_D\rangle,\ |\uparrow_B\downarrow_D\uparrow_C\rangle,\ \underbrace{|\uparrow_C\downarrow_B\uparrow_D\rangle}_{\exp\{-i\varphi_C-i\varphi_D\}},\ |\uparrow_C\downarrow_D\uparrow_B\rangle,\ \underbrace{|\uparrow_D\downarrow_B\uparrow_C\rangle}_{\exp\{-i\varphi_C-i\varphi_D\}},\ |\uparrow_D\downarrow_C\uparrow_B\rangle,\\
&|\uparrow_A\downarrow_C\uparrow_D\rangle,\ |\uparrow_A\downarrow_D\uparrow_C\rangle,\ \underbrace{|\uparrow_C\downarrow_A\uparrow_D\rangle}_{\exp\{-i\varphi_C-i\varphi_D\}},\ |\uparrow_C\downarrow_D\uparrow_A\rangle,\ \underbrace{|\uparrow_D\downarrow_A\uparrow_C\rangle}_{\exp\{-i\varphi_C-i\varphi_D\}},\ |\uparrow_D\downarrow_C\uparrow_A\rangle,\\
&|\uparrow_A\downarrow_B\uparrow_D\rangle,\ \underbrace{|\uparrow_A\downarrow_D\uparrow_B\rangle}_{\exp\{-i\varphi_A-i\varphi_B\}},\ |\uparrow_B\downarrow_A\uparrow_D\rangle,\ \underbrace{|\uparrow_B\downarrow_D\uparrow_A\rangle}_{\exp\{-i\varphi_A-i\varphi_B\}},\ |\uparrow_D\downarrow_A\uparrow_B\rangle,\ |\uparrow_D\downarrow_B\uparrow_A\rangle,\\
&|\uparrow_A\downarrow_B\uparrow_C\rangle,\ \underbrace{|\uparrow_A\downarrow_C\uparrow_B\rangle}_{\exp\{-i\varphi_A-i\varphi_B\}},\ |\uparrow_B\downarrow_A\uparrow_C\rangle,\ \underbrace{|\uparrow_B\downarrow_C\uparrow_A\rangle}_{\exp\{-i\varphi_A-i\varphi_B\}},\ |\uparrow_C\downarrow_A\uparrow_B\rangle,\ |\uparrow_C\downarrow_B\uparrow_A\rangle.
\end{aligned}
\tag{6.7}
$$

The order of arrows from left to right within a ket (or bra) specifies the sequence of pulse actions, which need not coincide with the chronology of arrival times as the pulses, although ultrashort, are of nonzero duration $\sim\sigma$ on the order of at most a few tens of femtoseconds.

Among the 24 three-pulse amplitudes listed in eqn (6.7) are eight which cannot contribute to the measured WPI signal and need not be developed further. As indicated with under-braces, these amplitudes will be seen, by arguments soon to be given, to carry nonsurviving optical phase factors. Overlapped with the appropriate term from eqn (6.6), each of them would participate in a contribution to the quadrilinear e-state population proportional to $\exp\{\pm i(\varphi_A+\varphi_B)\}$ and/or $\exp\{\pm i(\varphi_C+\varphi_D)\}$. These phase factors distribute themselves on a unit circle centered at the origin of the complex plane and hence average to "zero" relative to the phase-stable contributions over the many laser shots needed to accumulate a WPI signal at a given set of inter-pulse delays and controlled values of φ_{BA} and φ_{DC}.

Visualize a "nonsurviving" signal contribution proportional to $\exp\{i\varphi\}$ and collected over N laser shots as a random walk in the complex plane, for each successive step in which φ varies randomly within the range from zero to 2π. Argue that its net size grows as $\sqrt{N}$ and is therefore negligible compared to the phase-stable signal, which grows as N.

As a result, we need only find formulas for the sixteen remaining three-pulse amplitudes. Together with the complementary one-pulse amplitudes, these produce sixteen wave-packet overlaps contributing a measured 2D-WPI signal:

$$
\begin{aligned}
P = 2\mathrm{Re}\{&\langle\uparrow_A|\uparrow_B\downarrow_C\uparrow_D\rangle + \langle\uparrow_A|\uparrow_D\downarrow_C\uparrow_B\rangle + \langle\uparrow_A\downarrow_D\uparrow_C|\uparrow_B\rangle + \langle\uparrow_C\downarrow_D\uparrow_A|\uparrow_B\rangle\\
&+ \langle\uparrow_C|\uparrow_B\downarrow_A\uparrow_D\rangle + \langle\uparrow_C|\uparrow_D\downarrow_A\uparrow_B\rangle + \langle\uparrow_A\downarrow_B\uparrow_C|\uparrow_D\rangle + \langle\uparrow_C\downarrow_B\uparrow_A|\uparrow_D\rangle\\
&+ \langle\uparrow_A|\uparrow_B\downarrow_D\uparrow_C\rangle + \langle\uparrow_A|\uparrow_C\downarrow_D\uparrow_B\rangle + \langle\uparrow_A\downarrow_C\uparrow_D|\uparrow_B\rangle + \langle\uparrow_D\downarrow_C\uparrow_A|\uparrow_B\rangle\\
&+ \langle\uparrow_A\downarrow_B\uparrow_D|\uparrow_C\rangle + \langle\uparrow_D\downarrow_B\uparrow_A|\uparrow_C\rangle + \langle\uparrow_D|\uparrow_B\downarrow_A\uparrow_C\rangle + \langle\uparrow_D|\uparrow_C\downarrow_A\uparrow_B\rangle\}.
\end{aligned}
\tag{6.8}
$$

For later notational convenience we have chosen, between each quadrilinear overlap and its complex conjugate, the one in which the A pulse drives an absorptive transition in the bra or an emissive transition in the ket. Additionally, the sixteen inner products appearing in P have been grouped so that the first eight of them are those proportional to $\exp\{-i\varphi_{DC}-i\varphi_{BA}\}$ and the remaining eight are proportional to $\exp\{i\varphi_{DC}-i\varphi_{BA}\}$. Separating terms of sum and difference phasing, which we'll see shortly can be isolated experimentally, leads to $P = P^{(s)} + P^{(d)}$ with

$$P^{(s)} = 2\mathrm{Re}\{\underbrace{\langle\uparrow_A \mid \uparrow_B\downarrow_C\uparrow_D\rangle}_{\mathrm{SE}} + \langle\uparrow_A \mid \uparrow_D\downarrow_C\uparrow_B\rangle + \langle\uparrow_A\downarrow_D\uparrow_C \mid \uparrow_B\rangle + \langle\uparrow_C\downarrow_D\uparrow_A \mid \uparrow_B\rangle$$
$$+ \langle\uparrow_C \mid \uparrow_B\downarrow_A\uparrow_D\rangle + \langle\uparrow_C \mid \uparrow_D\downarrow_A\uparrow_B\rangle + \underbrace{\langle\uparrow_A\downarrow_B\uparrow_C \mid \uparrow_D\rangle}_{\mathrm{GSB}} + \langle\uparrow_C\downarrow_B\uparrow_A \mid \uparrow_D\rangle\} \tag{6.9}$$

and

$$P^{(d)} = 2\mathrm{Re}\{\langle\uparrow_A \mid \uparrow_B\downarrow_D\uparrow_C\rangle + \underbrace{\langle\uparrow_A \mid \uparrow_C\downarrow_D\uparrow_B\rangle}_{\mathrm{GSB}'} + \underbrace{\langle\uparrow_A\downarrow_C\uparrow_D \mid \uparrow_B\rangle}_{\mathrm{SE}} + \langle\uparrow_D\downarrow_C\uparrow_A \mid \uparrow_B\rangle$$
$$+ \underbrace{\langle\uparrow_A\downarrow_B\uparrow_D \mid \uparrow_C\rangle}_{\mathrm{GSB}} + \langle\uparrow_D\downarrow_B\uparrow_A \mid \uparrow_C\rangle + \langle\uparrow_D \mid \uparrow_B\downarrow_A\uparrow_C\rangle + \underbrace{\langle\uparrow_D \mid \uparrow_C\downarrow_A\uparrow_B\rangle}_{\mathrm{SE}'}\}. \tag{6.10}$$

Among the sum- and difference-phased overlaps, those in which the pulses act "in order" in the three-pulse bra or ket have been ascribed physical interpretations corresponding to ground-state bleach or stimulated emission. But by their construction, all of the overlaps contributing to the quadrilinear *e*-state population of a given phase signature can be generated from any one of them by some combination of phase-preserving exchanges of pulse labels: $A \leftrightarrow B$ and $C \leftrightarrow D$, $A \leftrightarrow C$, and $B \leftrightarrow D$ within $P^{(s)}$; $A \leftrightarrow B$ and $C \leftrightarrow D$, $A \leftrightarrow D$, and $B \leftrightarrow C$ within $P^{(d)}$. It is interesting to note that in fluorescence-detected WPI (in contrast to FWM, as we shall see shortly), none of the pulses plays a privileged role. Each one-pulse bra (or ket) makes contributing overlaps with all the kets (or bras) in which the three other pulses participate, regardless of their order of action (except for the fact that only phase-stable overlaps can make non-negligible signal contributions).

Provide physical justification for designating as SE or GSB specific overlaps in eqns (6.9) and (6.10). Verify the assertion that phase-preserving exchanges of pulse labels enable the generation of all the contributions to either $P^{(s)}$ or $P^{(d)}$ from any single term therein.

The various quadrilinear overlaps differ in their dependence on the interpulse delays, and it is through this delay dependence that WPI data provide information on the coherent dynamics of a molecular system. For bookkeeping purposes—and without any sacrifice of experimental data—we shall require that $t_A \leq t_B$, $t_C \leq t_D$, and $(t_A+t_B)/2 \leq (t_C+t_D)/2$. Assuming that all four pulses have identical envelopes, any data obtained with t_B less than t_A would coincide with those within the prescribed range having t_A and t_B interchanged and φ_{BA} changed in sign; and similarly for data with t_D less than t_C. Data with the midpoint of t_C and t_D less than that of t_A and t_B exists in the prescribed range with t_A interchanged with t_C, t_B interchanged with t_D, and φ_{BA} swapped with φ_{DC}. An exhaustive range of interpulse delays thus spans a three-dimensional space with t_{BA} and t_{DC} greater than or equal to zero and t_{CB} greater than or equal to $-(t_{DC}+t_{BA})/2$. Notice that under these restrictions, t_B can

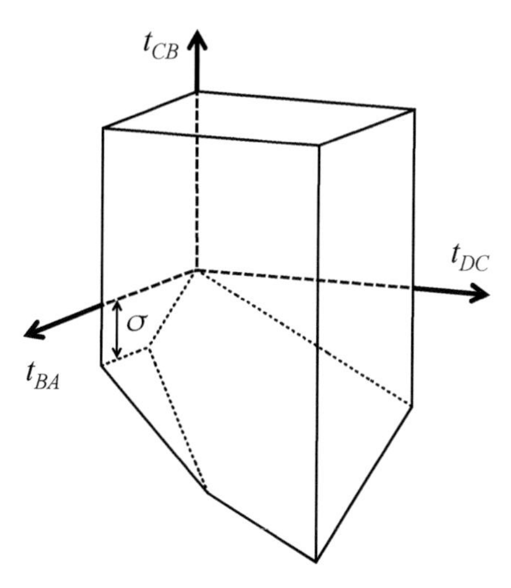

Fig. 6.1 Three-dimensional region of time-delay parameter space in which the ground-state-bleach contribution to the difference-phased 2D-WPI signal may be nonvanishing. The region shown is to be extended to arbitrarily large, positive values of t_{BA} and t_{DC}, and to arbitrarily large positive and negative values of t_{CB} (while maintaining t_{CB} greater than both $-(t_{DC}+t_{BA})/2$ and $-\sigma - t_{DC}$ in the latter case). The SE overlap may be nonzero in the temporal region generated from that depicted here by reflection through the $t_{DC} = t_{BA}$ plane. Reprinted by permission from Springer Nature: J. A. Cina and A. J. Kiessling in *Coherent Multidimensional Spectroscopy*, edited by M. Cho (Springer Nature, Singapore, 2019).

be before, between, or after t_C and t_D, and t_A can come before or after t_C, provided $t_{DC} + 2t_{CB} + t_{BA}$ remains nonnegative.

We can show that over three-dimensional regions of the parameter space of inter-pulse delays, the wave-packet overlaps labeled GSB and SE in $P^{(s)}$ and $P^{(d)}$, along with those labeled GSB′ and SE′ in $P^{(d)}$, are not excluded from making contributions to the 2D-WPI signal. In contrast, the other quadrilinear overlaps appearing in eqns (6.9) and (6.10) are confined to quasi two- or zero-dimensional regions (one or three of whose delay ranges, respectively, shrink to zero with decreasing pulse duration) and make correspondingly less globally important contributions to the WPI signal.

Let's consider $\langle \uparrow_A \downarrow_B \uparrow_D \mid \uparrow_C \rangle$, the GSB overlap of $P^{(d)}$. Since the D pulse acts after the B pulse in this term, the overlap vanishes if t_D precedes t_B by more than approximately the pulse duration, whence $t_D - t_B > -\sigma$, or $t_{DC} + t_{CB} + \sigma > 0$. Together with the general restriction that $t_{DC} + 2t_{CB} + t_{BA}$ be nonnegative, this condition confines the overlap to the 3D temporal region plotted in Fig. 6.1. A similar argument for the SE overlap of $P^{(d)}$, $\langle \uparrow_A \downarrow_C \uparrow_D \mid \uparrow_B \rangle$, shows that this term can only exist within the 3D region where $t_{CB} + t_{BA} + \sigma > 0$ and $t_{DC} + 2t_{CB} + t_{BA} > 0$, which corresponds to the mirror image through the vertical plane $t_{DC} = t_{BA}$ of the volume shown in Fig. 6.1.

For the GSB′ and SE′ overlaps also, $\langle \uparrow_A \mid \uparrow_C \downarrow_D \uparrow_B \rangle$ and $\langle \uparrow_D \mid \uparrow_C \downarrow_A \uparrow_B \rangle$, respectively, there are 3D regions of delay space within which signal contributions can pos-

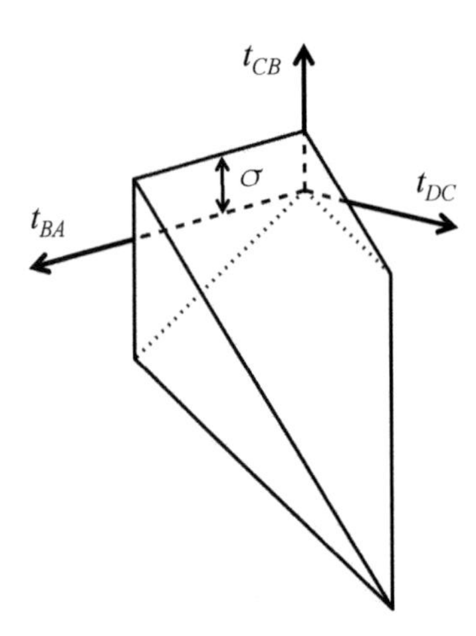

Fig. 6.2 Delay region in which the GSB′ contribution to the difference-phased 2D-WPI signal may not vanish. The pictured volume is to be extended to arbitrarily large, positive values of t_{BA} and t_{DC}, while maintaining t_{CB} greater than $-(t_{DC} + t_{BA})/2$ and less than $\sigma - t_{DC}$. The region of possibly nonvanishing SE′ is the mirror image of this one through the vertical $t_{DC} = t_{BA}$ plane. Reprinted by permission from Springer Nature: J. A. Cina and A. J. Kiessling in *Coherent Multidimensional Spectroscopy*, edited by M. Cho (Springer Nature, Singapore, 2019).

sibly occur. Since pulse D acts before pulse B in the first of these overlaps, their arrival times must obey $t_D - t_B < \sigma$, and GSB′ is therefore confined to the shared range of $t_{DC} + t_{CB} < \sigma$ and $t_{DC} + 2t_{CB} + t_{BA} > 0$ plotted in Fig. 6.2. For the SE′ overlap, on the other hand, pulse C must act before pulse A, whence $t_C - t_A < \sigma$. A nonvanishing SE′ contribution to $P^{(d)}$ is therefore possible when $t_{CB} + t_{BA} < \sigma$ and $t_{DC} + 2t_{CB} + t_{BA} > 0$. This region mirrors the volume shown in Fig. 6.2 through the $t_{DC} = t_{BA}$ plane.

In the overlap $\langle \uparrow_A \mid \uparrow_B \downarrow_D \uparrow_C \rangle$, the D pulse acts before the C pulse, despite the fact that by definition the arrival time t_D follows t_C; it must therefore be that $t_{DC} < \sigma$. Since the B pulse also acts before the C pulse in this overlap, but t_C can precede t_B in the specified range of unique interpulse delays, $t_B - t_C$ has to be less than the pulse length, or $t_{CB} > -\sigma$ is required as well. Together with $t_{DC} + 2t_{CB} + t_{BA} > 0$, these conditions restrict $\langle \uparrow_A \mid \uparrow_B \downarrow_D \uparrow_C \rangle$ to the quasi-2D time-delay region illustrated in Fig. 6.3. The delay region for $\langle \uparrow_D \mid \uparrow_B \downarrow_A \uparrow_C \rangle$ can be obtained by reflecting this slab in the vertical plane $t_{DC} = t_{BA}$.

The two remaining overlaps in $P^{(d)}$ are limited to quasi zero-dimensional delay regions. Since pulse B and D both act before pulse A in $\langle \uparrow_D \downarrow_B \uparrow_A \mid \uparrow_C \rangle$, the overlap cannot contribute unless t_{BA} and $t_D - t_A = t_{DC} + t_{CB} + t_{BA}$ are less than about σ. These conditions, together with the restriction to nonnegative $t_{DC} + 2t_{CB} + t_{BA}$, limit this overlap to the region shown in Fig. 6.4. The mirror image of this region similarly confines $\langle \uparrow_D \downarrow_C \uparrow_A \mid \uparrow_B \rangle$.

It is pertinent to recall that—despite their differing phase signatures, pulse orderings, and confining regions of time-delay space—all sixteen overlaps appearing in

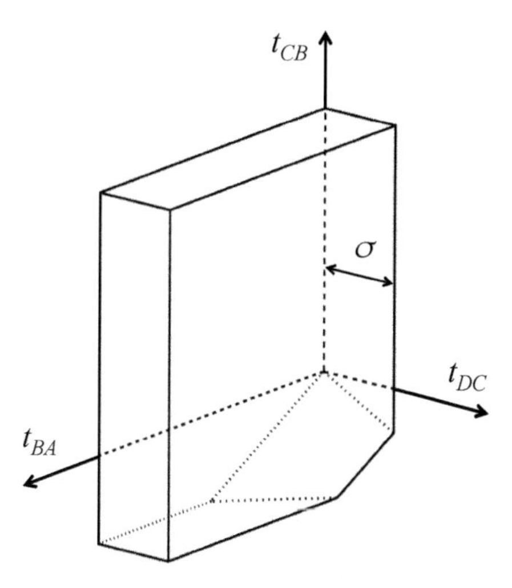

Fig. 6.3 Quasi-2D time-delay region in which the overlap $\langle \uparrow_A \mid \uparrow_B \downarrow_D \uparrow_C \rangle$ may contribute to the 2D-WPI signal; it is to be extended to arbitrarily large, positive values of t_{BA} and t_{CB}. Reflection in the plane $t_{DC} = t_{BA}$ yields the slab within which $\langle \uparrow_D \mid \uparrow_B \downarrow_A \uparrow_C \rangle$ may take nonzero values. Reprinted by permission from Springer Nature: J. A. Cina and A. J. Kiessling in *Coherent Multidimensional Spectroscopy*, edited by M. Cho (Springer Nature, Singapore, 2019).

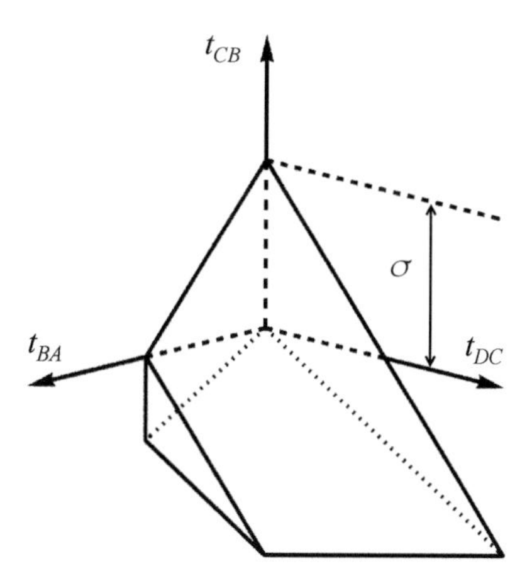

Fig. 6.4 Delay region where $\langle \uparrow_D \downarrow_B \uparrow_A \mid \uparrow_C \rangle$ may take nonnegligible values. This region is quasi zero-dimensional because it shrinks to nonexistence in all three directions as the pulse duration gets shorter and shorter. The overlap $\langle \uparrow_D \downarrow_C \uparrow_A \mid \uparrow_B \rangle$ may exist only inside a region mirroring this one through the $t_{DC} = t_{BA}$ plane. Reprinted by permission from Springer Nature: J. A. Cina and A. J. Kiessling in *Coherent Multidimensional Spectroscopy*, edited by M. Cho (Springer Nature, Singapore, 2019).

eqns (6.9) and (6.10) are essentially the same from the molecule's point of view: Each term records a contribution to the e-state population resulting from the interference between an amplitude generated by the action of a single resonant electric-field interaction and another generated by an excitation-deexcitation-reexcitation process driven by three sequential interactions with resonant fields. The pulse labels attached to the participating fields don't evince themselves in a single population-generation process, but only in the way the measured quadrilinear contribution to the e-state population varies with the experimentally specified interpulse delays.

Sketch of the region of $\{t_{BA}, t_{DC}, t_{CB}\}$ space, with t_{BA} and t_{DC} nonnegative and $t_{CB} > -\frac{1}{2}(t_{BA} + t_{DC})$, to which each overlap in $P^{(s)}$ of eqn (6.9) is confined.

Let us next discover how—as claimed earlier—it is possible experimentally to isolate $P^{(s)}$ and $P^{(d)}$. In fact, the complex-valued sums of wave-packet overlaps inside the braces of eqns (6.9) and (6.10) can themselves be separately obtained. For letting $P^{(s)} = 2\mathrm{Re}\{\xi_s \exp(-i\varphi_s)\}$ and $P^{(d)} = 2\mathrm{Re}\{\xi_d \exp(-i\varphi_d)\}$, where $\varphi_s = \varphi_{BA} + \varphi_{DC}$, $\varphi_d = \varphi_{BA} - \varphi_{DC}$, $\xi_s = \xi_s' + i\xi_s''$, and $\xi_d = \xi_d' + i\xi_d''$, one can isolate the real and imaginary parts of ξ_s and ξ_d by combining measurements of $P(\varphi_s, \varphi_d) = P^{(s)}(\varphi_s) + P^{(d)}(\varphi_d)$ with different optical phase shifts. We could for instance adopt the choices

$$\begin{aligned} P(0,0) &= 2(\xi_s' + \xi_d')\,, \\ P(0,\pi) &= 2(\xi_s' - \xi_d')\,, \\ P(\tfrac{\pi}{2},\tfrac{\pi}{2}) &= 2(\xi_s'' + \xi_d'')\,, \\ P(\tfrac{\pi}{2},-\tfrac{\pi}{2}) &= 2(\xi_s'' - \xi_d'')\,, \end{aligned} \tag{6.11}$$

which lead to

$$\xi_s = \frac{1}{4}\{P(0,0) + P(0,\pi) + iP(\tfrac{\pi}{2},\tfrac{\pi}{2}) + iP(\tfrac{\pi}{2},-\tfrac{\pi}{2})\}\,, \tag{6.12}$$

and

$$\xi_d = \frac{1}{4}\{P(0,0) - P(0,\pi) + iP(\tfrac{\pi}{2},\tfrac{\pi}{2}) - iP(\tfrac{\pi}{2},-\tfrac{\pi}{2})\}\,. \tag{6.13}$$

Once measured in this way, these complex quantities can be used to reconstruct the sought-after sums of overlaps as

$$\begin{aligned} \xi_s e^{-i\varphi_{BA} - i\varphi_{DC}} = &\langle \uparrow_A | \uparrow_B \downarrow_C \uparrow_D \rangle + \langle \uparrow_A | \uparrow_D \downarrow_C \uparrow_B \rangle + \langle \uparrow_A \downarrow_D \uparrow_C | \uparrow_B \rangle + \langle \uparrow_C \downarrow_D \uparrow_A | \uparrow_B \rangle \\ &+ \langle \uparrow_C | \uparrow_B \downarrow_A \uparrow_D \rangle + \langle \uparrow_C | \uparrow_D \downarrow_A \uparrow_B \rangle + \langle \uparrow_A \downarrow_B \uparrow_C | \uparrow_D \rangle + \langle \uparrow_C \downarrow_B \uparrow_A | \uparrow_D \rangle\,, \end{aligned} \tag{6.14}$$

and

$$\begin{aligned} \xi_d e^{-i\varphi_{BA} + i\varphi_{DC}} = &\langle \uparrow_A | \uparrow_B \downarrow_D \uparrow_C \rangle + \langle \uparrow_A | \uparrow_C \downarrow_D \uparrow_B \rangle + \langle \uparrow_A \downarrow_C \uparrow_D | \uparrow_B \rangle + \langle \uparrow_D \downarrow_C \uparrow_A | \uparrow_B \rangle \\ &+ \langle \uparrow_A \downarrow_B \uparrow_D | \uparrow_C \rangle + \langle \uparrow_D \downarrow_B \uparrow_A | \uparrow_C \rangle + \langle \uparrow_D | \uparrow_B \downarrow_A \uparrow_C \rangle + \langle \uparrow_D | \uparrow_C \downarrow_A \uparrow_B \rangle\,. \end{aligned} \tag{6.15}$$

Since the quantities in eqns (6.14) and (6.15) are sums of several overlaps, it might appear doubtful whether WPI spectroscopy can yield readily interpretable dynamical information. But six of the overlaps appearing eqn (6.14) and four of those in eqn

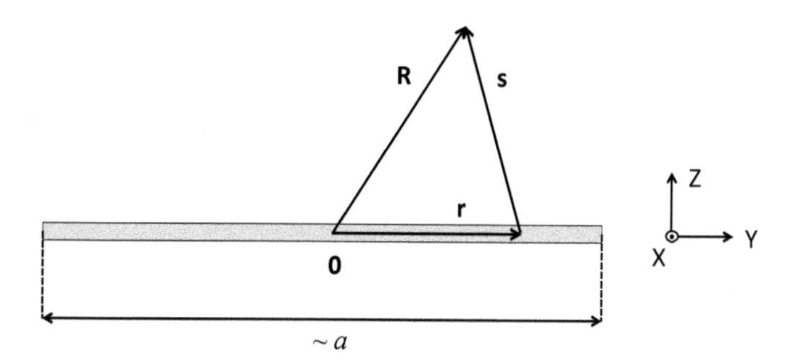

Fig. 6.5 Laser-illuminated sample layer in FWM experiment shown edge-on.

(6.15) vanish over much of the prescribed range of $\{t_{BA}, t_{DC}, t_{CB}\}$, leaving only the SE and GSB overlaps in the former, and the SE, GSB, SE′, and GSB′ in the latter.[6]

It is noteworthy that the basic expressions for the fluorescence-detected 2D-WPI signal do not assume some uniform density of resonant species within a spatially extended sample. Measurements of this kind could be carried out by accumulating 2D-WPI data from a single molecule whose fluorescence is repeatedly excited and detected at each combination of interpulse delays and phase shifts.[7] We will see presently that this capability distinguishes fluorescence-detected WPI from heterodyned FWM, as the latter method involves the generation of a signal beam emanating from an extended laser spot containing many chromophores and its interference with a co-propagating local-oscillator beam.

6.2.2 Heterodyne-detected four-wave mixing

The theoretical set-up for for FWM is somewhat more involved. The FWM signal field at some point $\mathbf{R}$ outside the sample is the superposition of electric fields generated by the oscillating trilinear dipole moments $\mathbf{m}_{ABC}$ of all the laser-illuminated chromophores. In order to illustrate the process of signal-beam generation, it is sufficient to consider a quasi two-dimensional sample thinner than the relevant optical wavelengths.[8] The source dipoles lie at molecular locations $\mathbf{r}$ within this sample layer, and the vector from a given molecule to the field point is $\mathbf{s} \equiv \mathbf{R} - \mathbf{r} = s\mathbf{n}$, as sketched in Fig. 6.5.

The trilinearly induced dipole moment of a chromophore within this sample in principle comprises twelve independent contributions:

[6]An effective scheme for isolating wave-packet overlaps contributing to short-pulse optical spectroscopy signals using *phase cycling* by acousto-optic modulation was reported in P. F. Tekavec, T. R. Dyke, and A. H. Marcus, "Wave packet interferometry and quantum state reconstruction by acousto-optic phase modulation," J. Chem. Phys. **125**, 194303/1–19 (2006).

[7]A. H. Marcus, private communication (March 2020); S. Mueller, S. Draeger, X. Ma, M. Hensen, T. Kenneweg, W. Pfeiffer, and T. Brixner, "Fluorescence-detected two-quantum and one-quantum-two-quantum 2D electronic spectroscopy," J. Phys. Chem. Lett. **9**, 1964–1969 (2018).

[8]For such an illustrative sample, despite its spatially extended nature, macroscopic propagation effects related for instance to the frequency dependence of an effective index of refraction need not be explicitly incorporated. While they will not be addressed here, propagation effects in FWM must often be accounted for in practice; see D. M. Jonas, "Two-dimensional femtosecond spectroscopy," Annu. Rev. Phys. Chem. **54**, 425–463 (2003).

$$\begin{aligned}\mathbf{m}_{ABC} = 2\mathrm{Re}\{&\langle\uparrow_A\downarrow_B\uparrow_C|\hat{\mathbf{m}}|0\rangle + \underbrace{\langle\uparrow_A\downarrow_C\uparrow_B|\hat{\mathbf{m}}|0\rangle}_{\exp\{i\varphi_A+i\varphi_B\}} + \langle 0|\hat{\mathbf{m}}|\uparrow_B\downarrow_A\uparrow_C\rangle \\ &+ \underbrace{\langle\uparrow_B\downarrow_C\uparrow_A|\hat{\mathbf{m}}|0\rangle}_{\exp\{i\varphi_A+i\varphi_B\}} + \langle 0|\hat{\mathbf{m}}|\uparrow_C\downarrow_A\uparrow_B\rangle + \langle\uparrow_C\downarrow_B\uparrow_A|\hat{\mathbf{m}}|0\rangle \\ &+ \langle\uparrow_A|\hat{\mathbf{m}}|\uparrow_B\downarrow_C\rangle + \underbrace{\langle\uparrow_A|\hat{\mathbf{m}}|\uparrow_C\downarrow_B\rangle}_{\exp\{i\varphi_A+i\varphi_B\}} + \langle\uparrow_A\downarrow_C|\hat{\mathbf{m}}|\uparrow_B\rangle \\ &+ \underbrace{\langle\uparrow_B|\hat{\mathbf{m}}|\uparrow_C\downarrow_A\rangle}_{\exp\{i\varphi_A+i\varphi_B\}} + \langle\uparrow_A\downarrow_B|\hat{\mathbf{m}}|\uparrow_C\rangle + \langle\uparrow_C|\hat{\mathbf{m}}|\uparrow_B\downarrow_A\rangle\}; \qquad (6.16)\end{aligned}$$

but the four underbraced matrix elements carry uncontrolled optical phase factors, and hence don't measurably contribute to the signal field. We can further anticipate the fact that four of the remaining dipole elements will contribute to the FWM signal with an optical phase signature $\exp\{-i\varphi_{DC} - i\varphi_{BA}\}$ and four with $\exp\{i\varphi_{DC} - i\varphi_{BA}\}$, and therefore write $\mathbf{m}_{ABC} = \mathbf{m}^{(s)} + \mathbf{m}^{(d)}$, where

$$\begin{aligned}\mathbf{m}^{(s)} = 2\mathrm{Re}\{&\langle\uparrow_A\downarrow_B\uparrow_C|\hat{\mathbf{m}}|0\rangle + \langle\uparrow_C\downarrow_B\uparrow_A|\hat{\mathbf{m}}|0\rangle \\ &+ \langle\uparrow_A|\hat{\mathbf{m}}|\uparrow_B\downarrow_C\rangle + \langle\uparrow_C|\hat{\mathbf{m}}|\uparrow_B\downarrow_A\rangle\}, \qquad (6.17)\end{aligned}$$

and

$$\begin{aligned}\mathbf{m}^{(d)} = 2\mathrm{Re}\{&\langle 0|\hat{\mathbf{m}}|\uparrow_B\downarrow_A\uparrow_C\rangle + \langle 0|\hat{\mathbf{m}}|\uparrow_C\downarrow_A\uparrow_B\rangle \\ &+ \langle\uparrow_A\downarrow_C|\hat{\mathbf{m}}|\uparrow_B\rangle + \langle\uparrow_A\downarrow_B|\hat{\mathbf{m}}|\uparrow_C\rangle\}. \qquad (6.18)\end{aligned}$$

Evaluating eqns (6.17) and (6.18) requires those one- and three-pulse kets listed in eqns (6.6) and (6.7) which do not involve the D pulse. The two-pulse amplitudes $|\uparrow_A\downarrow_B\rangle$, $|\uparrow_B\downarrow_A\rangle$, $|\uparrow_A\downarrow_C\rangle$, and $|\uparrow_B\downarrow_C\rangle$, along with the unperturbed ket $|0\rangle$, are also needed.

> Justify the expression (6.16) for the trilinear dipole moment as a consequence of the quantum mechanical superposition principle by considering the time-dependent state of the system as a sum of terms of zeroth, first, second, and third order in the field strengths, writing out the expectation value of the dipole moment operator in this state, and isolating the trilinear portion.

Before deriving the requisite zero- through three-pulse kets, let's develop an expression for the FWM signal itself. The net trilinear signal field generated in the far field of all the source dipoles—further than the optical wavelength from the sample plane, but not necessarily at a distance larger than the sample size, which corresponds to the diameter $\sim a$ of the laser spot—is given by the basic electrodynamical formula

$$\begin{aligned}\boldsymbol{\mathcal{E}}_{ABC}(t) &= \sum_{\mathbf{r}} \frac{1}{c^2 s}\left[\ddot{\mathbf{m}}_{ABC}\left(t - \frac{s}{c}\right) \times \mathbf{n}\right] \times \mathbf{n} \\ &= -\sum_{\mathbf{r}} \frac{1}{c^2 s}(\mathbf{1} - \mathbf{nn}) \cdot \ddot{\mathbf{m}}_{ABC}\left(t - \frac{s}{c}\right). \qquad (6.19)\end{aligned}$$

Here, s/c is the travel time of light from the molecular location $\mathbf{r}$ to the field point $\mathbf{R}$ (see Fig. 6.5). The signal field $\boldsymbol{\mathcal{E}}_{ABC}$ is superposed with the local oscillator $\mathbf{E}_D$, and difference detection measures the change in energy of the LO beam due to electromagnetic interference with the signal beam:

$$\Delta\mathcal{U} = \frac{1}{4\pi}\int d^3R\{[\mathbf{E}_D(t) + \boldsymbol{\mathcal{E}}_{ABC}(t)]^2 - E_D^2(t)]\}\,. \tag{6.20}$$

The observation time $t > t_D$ is late enough that the LO pulse has entirely passed the sample. For a given set of interpulse delays, $\Delta\mathcal{U}$ remains independent of the observation time thereafter; while it is convenient to calculate this quantity when the trailing edge of local-oscillator field is several wavelengths beyond the sample, it is typically measured *much* later than t_D and remotely from the sample at a distance large compared to the size of the laser spot. Because the signal field is weak, the formula for the FWM signal reduces to

$$\Delta\mathcal{U} \cong \frac{1}{2\pi}\int d^3R\,\mathbf{E}_D(t)\cdot\boldsymbol{\mathcal{E}}_{ABC}(t)\,. \tag{6.21}$$

Like the signal field, the LO field in eqns (6.20) and (6.21) is reckoned at the field point $\mathbf{R} = \mathbf{r} + \mathbf{s}$, rather than a location within the sample as in eqn (6.5). Using eqn (6.19) and integrating over the sample layer, we arrive at

$$\begin{aligned}\Delta\mathcal{U} = {} & \frac{\alpha}{2\pi c^2}\int d^2r\int d^3s\,\frac{1}{s}E_D f_D\left(t - t_D - \frac{\mathbf{r}\cdot\mathbf{n}_D}{c} - \frac{\mathbf{s}\cdot\mathbf{n}_D}{c}\right)\\ & \times\cos\left[\Omega\left(t - t_D - \frac{\mathbf{r}\cdot\mathbf{n}_D}{c} - \frac{\mathbf{s}\cdot\mathbf{n}_D}{c}\right) + \varphi_D\right]\left[(\mathbf{e}_D\cdot\mathbf{n})\mathbf{n} - \mathbf{e}_D\right]\cdot\ddot{\mathbf{m}}_{ABC}\left(t - \frac{s}{c}\right),\end{aligned} \tag{6.22}$$

where α is the number of chromophores per unit sample area.

> Draw a sketch illustrating the component isolation at work in the quantity $\mathbf{e}_D\cdot(\mathbf{1} - \mathbf{n}\mathbf{n})\cdot\ddot{\mathbf{m}}_{ABC}$.

Working from the general formula of eqn (6.22), we further analyze the case in which the LO beam propagates in the z-direction, with $\mathbf{n}_D = \mathbf{k}$ perpendicular to the sample plane and its polarization $\mathbf{e}_D = \mathbf{i}$ in the x-direction. We switch to spherical polar coordinates, and, for simplicity, assume that all molecular dipoles $\hat{\mathbf{m}} = m\mathbf{i}(|e\rangle\langle g| + |g\rangle\langle e|)$ are aligned with the x-axis.[9] Then

$$\begin{aligned}\Delta\mathcal{U} &= \frac{\alpha E_D}{2\pi c^2}\int d^2r\int_0^\infty ds\,s\,\ddot{m}_{ABC}\left(t - \frac{s}{c}\right)\int_0^\pi d\theta\,\sin\theta\,f_D\left(t - t_D - \frac{s}{c}\cos\theta\right)\\ &\quad\times\cos\left[\Omega\left(t - t_D - \frac{s}{c}\cos\theta\right) + \varphi_D\right]\int_0^{2\pi} d\phi\left(\sin^2\theta\cos^2\phi - 1\right)\\ &= -\frac{\alpha E_D}{2c^2}\int d^2r\int_0^\infty ds\,s\,\ddot{m}_{ABC}\left(t - \frac{s}{c}\right)I(s)\,,\end{aligned} \tag{6.23}$$

[9] A sample of nonzero thickness on which the LO pulse impinges at a near-normal angle could rigorously be regarded as consisting of numerous layers strictly perpendicular to the propagation direction of this incident beam.

where

$$I(s) = \int_0^\pi d\theta \sin\theta (1 + \cos^2\theta) \times f_D\left(t - t_D - \frac{s}{c}\cos\theta\right) \cos\left[\Omega\left(t - t_D - \frac{s}{c}\cos\theta\right) + \varphi_D\right]. \tag{6.24}$$

The substitution $u = t - t_D - \frac{s}{c}\cos\theta$ leads to

$$I(s) = \frac{c}{s}\int_{t-t_D-\frac{s}{c}}^{\infty} du \left[1 + \frac{c^2}{s^2}(t - t_D - u)^2\right] f_D(u) \cos(\Omega u + \varphi_D), \tag{6.25}$$

in which the upper limit has been extended from $t - t_D + \frac{s}{c}$ to infinity because $t - t_D$ exceeds σ by choice and s is nonnegative, while $f_D(u)$ vanishes for u greater than the pulse duration. Taking account of the slowly varying nature of the pulse envelope, the integral is found to be

$$\begin{aligned} I(s) &\cong f_D\left(t - t_D - \frac{s}{c}\right)\left\{-\left(\frac{2c}{s\Omega} - \frac{2c^3}{s^3\Omega^3}\right)\sin\left[\Omega\left(t - t_D - \frac{s}{c}\right) + \varphi_D\right]\right. \\ &\quad \left. + \frac{2c^2}{s^2\Omega^2}\cos\left[\Omega\left(t - t_D - \frac{s}{c}\right) + \varphi_D\right]\right\} \\ &\cong -\frac{2c}{s\Omega} f_D\left(t - t_D - \frac{s}{c}\right)\sin\left[\Omega\left(t - t_D - \frac{s}{c}\right) + \varphi_D\right]. \end{aligned} \tag{6.26}$$

The second relation in eqn (6.26) results from the fact the envelope of the D pulse localizes it to the spatial range $c(t - t_D - \sigma) < s < c(t - t_D + \sigma)$; for the pulse's trailing edge to lie in the far-field region of the molecular dipoles requires $2\pi c/\Omega < c(t - t_D - \sigma) < s$, which implies $s\Omega \gg c$. Inserting eqn (6.26) in eqn (6.23), changing the integration variable from s to $t' = t - \frac{s}{c}$ (only to drop the prime later), integrating by parts, and invoking the slowly varying-envelope approximation yield

$$\Delta\mathcal{U} = -\alpha E_D \Omega \int_{-\infty}^{\infty} dt\, f_D(t - t_D) \sin\left[\Omega(t - t_D) + \varphi_D\right] \int d^2r\, m_{ABC}(t). \tag{6.27}$$

The innermost integral, over the illuminated sample area, will later be shown to enforce a *wave-vector-matching* relation among the directions of propagation of the four laser beams.

In FWM as in 2D WPI, B is by definition the second-arriving pulse of the phase-related AB pair. From eqn (6.27), we see that $\Delta\mathcal{U}$ vanishes if the local-oscillator D pulse precedes the later arriving of the B and C pulses by more than about the pulse-length σ, as m_{ABC} will not yet have turned on when pulse D is present. In heterodyne-detected four-wave mixing, we therefore do not bother to require that t_C strictly precede t_D and also do not restrict $\frac{1}{2}(t_A + t_B)$ to values less than $\frac{1}{2}(t_C + t_D)$.

1. Verify both members of eqn (6.26).
2. Carry out the steps described to derive eqn (6.27).

6.2.3 Four-wave mixing by spectral interferometry

In one popular approach to heterodyne-detected four-wave mixing, the arrival time of the local oscillator is held fixed, *well in advance* of the other pulses. Instead of scanning t_{DC}, this *spectral interferometry* technique resolves the co-propagating transmitted signal and local-oscillator beams by frequency, in essence performing a Fourier transformation with respect to the intraprobe pulse-pair delay experimentally. Referring to the local oscillator in such a measurement as L, we have $t_L \ll t_A, t_B, t_C$,[10] and the observed quantity is

$$\delta u_{\bar{\omega}} = \frac{1}{2\pi} \int d^3R\, \mathbf{E}'_L(t) \cdot \boldsymbol{\mathcal{E}}'_{ABC}(t). \tag{6.28}$$

This expression differs from eqn (6.21) in having $\mathbf{E}_L$ and $\boldsymbol{\mathcal{E}}_{ABC}$ replaced by their spectrally filtered portions, signified by primes, which possess frequency components only within a narrow range about some $\bar{\omega}$.[11] In addition, it is assumed that the otherwise arbitrary observation time t is late enough that the spatially and temporally elongated filtered LO field resides entirely in the far-field region of the molecular dipoles.

The filtered LO field is a superposition of its Fourier components, with

$$\tilde{\mathbf{E}}_L(\omega > 0) \cong \mathbf{e}_L \frac{E_L}{2} e^{i\omega t_L(\mathbf{R}) - i\varphi_L} \tilde{f}_L(\omega - \Omega) \tag{6.29}$$

and $\tilde{\mathbf{E}}_L(\omega < 0) = \tilde{\mathbf{E}}^*_L(-\omega)$. Here $\tilde{f}_L(\xi) = \int_{-\infty}^{\infty} dt\, e^{i\xi t} f_L(t)$ is the Fourier transform of the local oscillator's envelope function. The spectrally filtered local oscillator then becomes

$$\begin{aligned}\mathbf{E}'_L(t) &= \int_{\bar{\omega}-\delta\omega/2}^{\bar{\omega}+\delta\omega/2} \frac{d\omega}{2\pi}\, e^{-i\omega t}\, \tilde{\mathbf{E}}_L(\omega) + \text{c.c.} \\ &= \mathbf{e}_L E_L \frac{\sin \frac{\delta\omega}{2}(t - t_L(\mathbf{R}))}{\pi(t - t_L(\mathbf{R}))}\ \mathrm{Re}\left\{\tilde{f}_L(\bar{\omega} - \Omega) e^{-i\bar{\omega}(t - t_L(\mathbf{R})) - i\varphi_L}\right\},\end{aligned} \tag{6.30}$$

which is seen to have a prolonged duration $\sigma' \cong 2\pi/\delta\omega \gg \sigma$ set by the spectral resolution $\delta\omega$. For the special case in which the pulse envelope is an even function of time, $\tilde{f}_L$ is real and eqn (6.30) reduces to

$$\mathbf{E}'_L(t) = \mathbf{e}_L E_L \frac{\sin \frac{\delta\omega}{2}(t - t_L(\mathbf{R}))}{\pi(t - t_L(\mathbf{R}))} \tilde{f}_L(\bar{\omega} - \Omega) \cos\left[\bar{\omega}(t - t_L(\mathbf{R})) + \varphi_L\right]. \tag{6.31}$$

The field-point dependence of the filtered local oscillator (6.30) is dominated its plane-wave factors $\exp\{\pm i\bar{\omega}\mathbf{n}_L \cdot \mathbf{R}/c\}$. In the spatial integral (6.28) over the region of electromagnetic interference, components of the signal field with complementary wave vectors will be selected. Since these also have complementary temporal oscillations at

[10]The LO is made to precede the other pulses in order to prevent its contamination by spectroscopically uninformative effects of those pulses on the host medium.

[11]Like the carrier frequency Ω, the observed frequency $\bar{\omega}$ is regarded as positive.

frequencies near $\bar{\omega}$, frequency filtration of the signal field is unnecessary and eqn (6.28) can equally well be written

$$\delta u_{\bar{\omega}} = \frac{1}{2\pi} \int d^3R\, \mathbf{E}'_L(t) \cdot \boldsymbol{\mathcal{E}}_{ABC}(t)\,. \tag{6.32}$$

Comparing eqn (6.32) with eqn (6.21) makes it is clear that converting this spatial integral to an integral over time leads to an expression analogous to eqn (6.27):

$$\delta u_{\bar{\omega}} = \alpha E_L \bar{\omega} \int_{-\infty}^{\infty} dt\, \frac{\sin\frac{\delta\omega}{2}(t-t_L)}{\pi(t-t_L)} \times \text{Im}\left\{\tilde{f}_L(\bar{\omega}-\Omega)e^{-i\bar{\omega}(t-t_L)-i\varphi_L}\right\} \int d^2r\, m_{ABC}(t)\,. \tag{6.33}$$

1. Determine $\tilde{f}(\xi)$ explicitly for the case of a truncated-cosine envelope, $f(t) = \cos(\pi t/2\sigma)$ for $-\sigma < t < \sigma$ and zero otherwise.
2. Convince yourself of eqn (6.33)'s veracity. What form does this signal expression take when the envelope of the L-pulse is an even function of time?

We next investigate the relationship between the spectral interferometry signal of eqn (6.33) and the heterodyne-detected FWM signal (6.27) as detected with an unfiltered, delayed local oscillator. Simply summing $\delta u_{\bar{\omega}}$ of over a full spectrum of detection frequencies (and recalling that $\bar{\omega}$ is nonnegative) gives

$$\sum_{\bar{\omega}\geq 0} \delta u_{\bar{\omega}} = \alpha E_L\, \text{Im} \int_{-\infty}^{\infty} dt \sum_{\bar{\omega}} \bar{\omega} \frac{\sin\frac{\delta\omega}{2}(t-t_L)}{\pi(t-t_L)} \times \tilde{f}_L(\bar{\omega}-\Omega)e^{-i\bar{\omega}(t-t_L)-i\varphi_L} \int d^2r\, m_{ABC}(t)\,. \tag{6.34}$$

Here, we can make the replacement

$$\frac{\sin\frac{\delta\omega}{2}(t-t_L)}{\pi(t-t_L)} = \int_{-\frac{\delta\omega}{2}}^{\frac{\delta\omega}{2}} \frac{d\omega''}{2\pi}\, e^{-i\omega''(t-t_L)}\,, \tag{6.35}$$

whence, with a change of integration variables,

$$\begin{aligned} \sum_{\bar{\omega}\geq 0} \delta u_{\bar{\omega}} &\cong \alpha E_L\, \text{Im} \int_{-\infty}^{\infty} dt \int_0^{\infty} \frac{d\omega}{2\pi}\, \omega \tilde{f}_L(\omega-\Omega)e^{-i\omega(t-t_L)-i\varphi_L} \int d^2r\, m_{ABC}(t) \\ &= -\alpha E_L \frac{\partial}{\partial t_L}\, \text{Re} \int_{-\infty}^{\infty} dt \int_0^{\infty} \frac{d\omega}{2\pi}\, \tilde{f}_L(\omega-\Omega)e^{-i\omega(t-t_L)-i\varphi_L} \int d^2r\, m_{ABC}(t)\,. \end{aligned} \tag{6.36}$$

The lower limit of the frequency integral can be extended to $-\infty$ because $\tilde{f}_L(\omega-\Omega)$ is peaked at $\omega \approx \Omega$. Noting that

$$\text{Re} \int_{-\infty}^{\infty} \frac{d\omega}{2\pi}\, \tilde{f}_L(\omega-\Omega)e^{-i\omega(t-t_L)-i\varphi_L} = f_L(t-t_L)\cos\left[\Omega(t-t_L)+\varphi_L\right] \tag{6.37}$$

leads finally to

$$\sum_{\bar{\omega}\geq 0} \delta u_{\bar{\omega}} = -\alpha E_L \frac{\partial}{\partial t_L} \int_{-\infty}^{\infty} dt\, f_L(t-t_L) \cos\left[\Omega(t-t_L)+\varphi_L\right] \int d^2r\, m_{ABC}(t)$$
$$\cong \Delta\mathcal{U}_{t_L}\,, \tag{6.38}$$

where the last step invokes the slowly varying nature of the L-pulse envelope. The arrival time of pulse D, indicated by the subscript and here taking the value t_L, precedes the turn-on of m_{ABC} in the present situation, so the third member of this equality *vanishes*.

A similar derivation can however provide a more useful result, as

$$2\sum_{\bar{\omega}\geq 0} \delta u_{\bar{\omega}} \cos\bar{\omega}\Delta t \cong \Delta\mathcal{U}_{t_L+\Delta t} + \Delta\mathcal{U}_{t_L-\Delta t} = \Delta\mathcal{U}_{t_L+\Delta t}\,, \tag{6.39}$$

for any positive delay increment obeying $\delta\omega\Delta t \ll 2\pi$. Hence, with sufficient resolution, spectral interferometry data with t_L earlier than t_A, t_B, and t_C can be used to synthesize heterodyne-detected FWM signals having the full range of relevant D-pulse arrivals, $t_L + \Delta t$, starting at $\max(t_B, t_C)$ less the pulse duration. The relation (6.39) can of course be inverted. Since the spectral resolution is assumed to be high,

$$\delta u_{\bar{\omega}} \cong \frac{1}{\delta\omega} \int_{\bar{\omega}-\frac{\delta\omega}{2}}^{\bar{\omega}+\frac{\delta\omega}{2}} d\zeta\, \delta u_\zeta\,, \tag{6.40}$$

from which we may construct an alternative form for eqn (6.39),

$$\Delta\mathcal{U}_{t_L+\Delta t} = \frac{2}{\delta\omega} \int_0^{\infty} d\zeta\, \delta u_\zeta \cos\zeta\Delta t\,. \tag{6.41}$$

Fourier transformation then gives

$$\delta u_\omega = \frac{\delta\omega}{2\pi} e^{-i\omega t_L} \int_{-\infty}^{\infty} d\tau\, e^{i\omega\tau} \Delta\mathcal{U}_\tau\,, \tag{6.42}$$

where the integrand vanishes for τ less than about $\max(t_B, t_C) - \sigma$.

Carry through the derivation of eqns (6.39) and (6.42).

6.2.4 Connecting 2D WPI with transient absorption

The basic formula (6.32) for the spectral interferometry-based FWM signal helps establish a connection between two-dimensional electronic spectroscopy and simpler, transient-absorption measurements (see Chapter 3). We consider the special circumstance in which the A and B pulses of a FWM experiment have identical arrival times, phases, envelopes, directions of propagation, and polarizations—in effect collapsing to a single pump pulse—while the C and L pulses share identical envelopes, propagation directions, and polarizations. Then the spectral interferometry-detected FWM formula,

$$\delta u_{\bar{\omega}} = \frac{1}{2\pi} \int d^3R \, \mathbf{E}'_L(t) \cdot \boldsymbol{\mathcal{E}}_{AAC}(t) \, . \tag{6.43}$$

resembles that for transient absorption (actually, transient transmission),

$$\delta s_{\bar{\omega}} = \frac{1}{2\pi} \int d^3R \, \mathbf{E}'_C(t) \cdot \boldsymbol{\mathcal{E}}_{AAC}(t) \, . \tag{6.44}$$

Equation (6.44) calculates the change in transmitted probe (C) intensity at frequency $\bar{\omega}$ due to the sample's prior interaction with a pump pulse (A). In $\delta u_{\bar{\omega}}$ and $\delta s_{\bar{\omega}}$, the common signal field $\boldsymbol{\mathcal{E}}_{AAC}$ interferes with the spectrally filtered fields,

$$\begin{aligned} \mathbf{E}'_L(t) = \mathbf{e}_C E_L & \frac{\sin \frac{\delta\omega}{2}(t - t_C(\mathbf{R}) - t_{LC})}{\pi(t - t_C(\mathbf{R}) - t_{LC})} \\ & \times \operatorname{Re}\left\{ \tilde{f}_C(\bar{\omega} - \Omega) e^{-i\bar{\omega}(t - t_C(\mathbf{R}) - t_{LC}) - i\varphi_C - i\varphi_{LC}} \right\}, \end{aligned} \tag{6.45}$$

from eqn (6.30), and

$$\mathbf{E}'_C(t) = \mathbf{e}_C E_C \frac{\sin \frac{\delta\omega}{2}(t - t_C(\mathbf{R}))}{\pi(t - t_C(\mathbf{R}))} \operatorname{Re}\left\{ \tilde{f}_C(\bar{\omega} - \Omega) e^{-i\bar{\omega}(t - t_C(\mathbf{R})) - i\varphi_C} \right\}, \tag{6.46}$$

respectively; the notation in eqn (6.45) reflects the position independence of the (negative) intrapulse-pair delay t_{LC}. The envelopes of the two filtered fields are peaked at different times $t_C(\mathbf{R}) + t_{LC}$ and $t_C(\mathbf{R})$; but since their common duration $\sim 2\pi/\delta\omega$ greatly exceeds $t_C - t_L$, this difference is negligible. Hence, by choosing an optical phase shift $\varphi_{LC} = \bar{\omega} t_{LC}$, the FWM signal at frequency $\bar{\omega}$ can be made essentially equivalent to the corresponding transient-absorption signal.

6.3 Quantum mechanical aspects

6.3.1 Time-dependent perturbation theory

Let's embark on the perturbative calculation of the one-, two-, and three-pulse kets needed to evaluate the various contributions to the quadrilinear e-state population and the trilinear electric dipole moment. At $t_0 \ll t_A(\mathbf{r})$, the initial state is taken to be

$$\begin{aligned} |\Psi(t_0)\rangle &= e^{-iH(t_0 - t_A(\mathbf{r}))/\hbar} |g\rangle |n_g\rangle \\ &= [\, t_0 - t_A(\mathbf{r}) \,] |g\rangle |n_g\rangle \\ &= |g\rangle [\, t_0 - t_A(\mathbf{r}) \,]_{gg} |n_g\rangle \\ &= |g\rangle |n_g\rangle e^{-i\omega_{n_g}(t_0 - t_A(\mathbf{r}))} \, , \end{aligned} \tag{6.47}$$

where we've adopted the streamlined notation $[\,t\,] = \exp\{-iHt/\hbar\}$ and $[\,t\,]_{jj} = \langle j | \exp\{-iHt/\hbar\} | j \rangle$, whence $[\,t\,]_{gg} = \exp\{-iH_g t/\hbar\}$ and $[\,t\,]_{ee} = \exp\{-iH_e t/\hbar - i\Lambda t\}$.

Upon switching to the interaction picture, $|\tilde{\Psi}(t)\rangle = [-t + t_A(\mathbf{r})]|\Psi(t)\rangle$, the initial condition becomes $|\tilde{\Psi}(t_0)\rangle = |g\rangle|n_g\rangle$ and the state evolves according to

$$i\hbar\frac{\partial|\tilde{\Psi}(t)\rangle}{\partial t} = \tilde{V}(t)|\tilde{\Psi}(t)\rangle\,, \tag{6.48}$$

with $\tilde{V}(t) \equiv [-t+t_A(\mathbf{r})]V(t)[t-t_A(\mathbf{r})]$. This interaction-picture Schrödinger equation has a formal solution,

$$|\tilde{\Psi}(t)\rangle = |g\rangle|n_g\rangle + \frac{1}{i\hbar}\int_{-\infty}^{t} d\tau\,\tilde{V}(\tau)|\tilde{\Psi}(\tau)\rangle\,; \tag{6.49}$$

the lower limit of integration can be extended to minus infinity because t_0 precedes all the laser pulses.

Iterating the formal solution gives the required third-order approximation,

$$\begin{aligned}|\tilde{\Psi}(t)\rangle \cong \Bigg\{1 + \frac{1}{i\hbar}\int_{-\infty}^{t} d\tau\,\tilde{V}(\tau) + \left(\frac{1}{i\hbar}\right)^2\int_{-\infty}^{t} d\tau\int_{-\infty}^{\tau} d\bar{\tau}\,\tilde{V}(\tau)\tilde{V}(\bar{\tau})\\ + \left(\frac{1}{i\hbar}\right)^3\int_{-\infty}^{t} d\tau\int_{-\infty}^{\tau} d\bar{\tau}\int_{-\infty}^{\bar{\tau}} d\bar{\bar{\tau}}\,\tilde{V}(\tau)\tilde{V}(\bar{\tau})\tilde{V}(\bar{\bar{\tau}})\Bigg\}|g\rangle|n_g\rangle\,.\end{aligned} \tag{6.50}$$

Writing V as a sum over pulses as in eqn (6.2) and reverting to the Schrödinger picture yield

$$\begin{aligned}|\Psi(t)\rangle \cong \Bigg\{&[t - t_A(\mathbf{r})]\\ &+ \frac{1}{i\hbar}\sum_I [t - t_I(\mathbf{r})]\int_{-\infty}^{t} d\tau\,[t_I(\mathbf{r}) - \tau]V_I(\tau)[\tau - t_I(\mathbf{r})][t_{IA}(\mathbf{r})]\\ &+ \left(\frac{1}{i\hbar}\right)^2\sum_{IJ}[t - t_J(\mathbf{r})]\int_{-\infty}^{t} d\tau\int_{-\infty}^{\tau} d\bar{\tau}\,[t_J(\mathbf{r}) - \tau]V_J(\tau)[\tau - t_J(\mathbf{r})][t_{JI}(\mathbf{r})]\\ &\qquad\times[t_I(\mathbf{r}) - \bar{\tau}]V_I(\bar{\tau})[\bar{\tau} - t_I(\mathbf{r})][t_{IA}(\mathbf{r})]\\ &+ \left(\frac{1}{i\hbar}\right)^3\sum_{IJK}[t - t_K(\mathbf{r})]\int_{-\infty}^{t} d\tau\int_{-\infty}^{\tau} d\bar{\tau}\int_{-\infty}^{\bar{\tau}} d\bar{\bar{\tau}}\,[t_K(\mathbf{r}) - \tau]V_K(\tau)[\tau - t_K(\mathbf{r})][t_{KJ}(\mathbf{r})]\\ &\times[t_J(\mathbf{r}) - \bar{\tau}]V_J(\bar{\tau})[\bar{\tau} - t_J(\mathbf{r})][t_{JI}(\mathbf{r})][t_I(\mathbf{r}) - \bar{\bar{\tau}}]V_I(\bar{\bar{\tau}})[\bar{\bar{\tau}} - t_I(\mathbf{r})][t_{IA}(\mathbf{r})]\Bigg\}|g\rangle|n_g\rangle\,.\end{aligned} \tag{6.51}$$

Note here that $[t_{JI}(\mathbf{r})]$ is the free time-evolution operator for the interpulse delay $t_{JI}(\mathbf{r}) = t_J(\mathbf{r}) - t_I(\mathbf{r})$. Equation (6.51) can be rendered less abstruse by introducing pulse propagators,

$$P_I(t;\tau) = \frac{E_I}{\hbar}\int_{-\infty}^{t} d\tau\,f_I(\tau)\cos(\Omega\tau + \varphi_I)[-\tau]\,\mathbf{e}_I\cdot\hat{\mathbf{m}}[\tau]\,. \tag{6.52}$$

In this notation, the first time argument of the pulse propagator is its upper limit of integration and the second variable names the quantity to be integrated over. Using eqn (6.52) and applying the appropriate shift in integration variable to the upper limit of any earlier pulse propagator allows us to rewrite the perturbed state as

$$|\Psi(t)\rangle = \Big\{ \big[t - t_A(\mathbf{r})\big] + i \sum_I \big[t - t_I(\mathbf{r})\big] P_I\big(t - t_I(\mathbf{r}); \tau\big)\big[t_{IA}(\mathbf{r})\big]$$

$$+ i^2 \sum_{IJ} \big[t - t_J(\mathbf{r})\big] P_J\big(t - t_J(\mathbf{r}); \tau\big)\big[t_{JI}(\mathbf{r})\big] P_I\big(\tau + t_{JI}(\mathbf{r}); \bar{\tau}\big)\big[t_{IA}(\mathbf{r})\big]$$

$$+ i^3 \sum_{IJK} \big[t - t_K(\mathbf{r})\big] P_K\big(t - t_K(\mathbf{r}); \tau\big)\big[t_{KJ}(\mathbf{r})\big] P_J\big(\tau + t_{KJ}(\mathbf{r}); \bar{\tau}\big)\big[t_{JI}(\mathbf{r})\big]$$

$$\times P_I\big(\bar{\tau} + t_{JI}(\mathbf{r}); \bar{\bar{\tau}}\big)\big[t_{IA}(\mathbf{r})\big]\Big\} |g\rangle |n_g\rangle . \qquad (6.53)$$

The definition (6.52) can be further developed by using the form of the dipole-moment operator (6.4) and making a rotating-wave approximation to obtain

$$P_I(t; \tau) = F_I e^{-i\varphi_I} |e\rangle\langle g| p^{(eg)}(t; \tau) + \text{H.c.}\,, \qquad (6.54)$$

where $F_I \equiv E_I \sigma(\mathbf{e}_I \cdot \mathbf{m})/2\hbar$. Reduced pulse propagators—which act only on nuclear degrees of freedom—are defined by

$$p_I^{(eg)}(t; \tau) = \int_{-\infty}^{t} \frac{d\tau}{\sigma} f_I(\tau) e^{-i\Omega\tau} [-\tau]_{ee} [\tau]_{gg}\,, \qquad (6.55)$$

and $\big(p_I^{(eg)}\big)^\dagger = p_I^{(ge)}$.[12]

Identify the contributions to P_I omitted by the rotating-wave approximation and explain their negligibility under resonant or near-resonant excitation.

6.3.2 Multilinear bras and kets

It is now a simple matter to abstract from $|\Psi\rangle$ in eqn (6.53) compact formulas for the zero- through three-pulse kets appearing in the quadrilinear e-state population and the trilinear dipole moment:

$$|0\rangle = |g\rangle [t - t_A(\mathbf{r})]_{gg} |n_g\rangle = |n_g\rangle e^{-i\omega_{n_g}(t - t_A(\mathbf{r}))}\,, \qquad (6.56)$$

$$|\uparrow_I\rangle = |e\rangle\, i F_I e^{-i\varphi_I} [t - t_I(\mathbf{r})]_{ee}\, p_I^{(eg)}(t - t_I(\mathbf{r}); \tau) [t_{IA}(\mathbf{r})]_{gg} |n_g\rangle\,, \qquad (6.57)$$

$$|\uparrow_I\downarrow_J\rangle = |g\rangle\, i^2 F_I F_J e^{i\varphi_{JI}} [t - t_J(\mathbf{r})]_{gg}\, p_J^{(ge)}(t - t_J(\mathbf{r}); \tau) [t_{JI}(\mathbf{r})]_{ee}$$

$$\times\, p_I^{(eg)}(\tau + t_{JI}(\mathbf{r}); \bar{\tau}) [t_{IA}(\mathbf{r})]_{gg} |n_g\rangle\,, \qquad (6.58)$$

and

[12] We have not specified a precise relationship between the pulse-duration parameter σ appearing in this definition and the pulse envelope, nor assigned different values of σ to different pulses; it is unnecessary to do either, because this parameter cancels out between F_I and p_I.

$$|\uparrow_I\downarrow_J\uparrow_K\rangle = |e\rangle\, i^3 F_I F_J F_K e^{i\varphi_{JI}-i\varphi_K}[\,t-t_K(\mathbf{r})\,]_{ee}\, p_K^{(eg)}(t-t_K(\mathbf{r});\tau)[\,t_{KJ}(\mathbf{r})\,]_{gg}$$
$$\times\, p_J^{(ge)}(\tau+t_{KJ}(\mathbf{r});\bar{\tau})[\,t_{JI}(\mathbf{r})\,]_{ee}\, p_I^{(eg)}(\bar{\tau}+t_{JI}(\mathbf{r});\bar{\bar{\tau}})[\,t_{IA}(\mathbf{r})\,]_{gg}|n_g\rangle\,. \quad (6.59)$$

In eqns (6.56)–(6.59), the nuclear wave packet accompanying the final electronic state ($|g\rangle$ for even-order components, $|e\rangle$ for odd) is economically described by alternating episodes of time-evolution under the appropriate nuclear Hamiltonian for a nominal interpulse delay and instantaneous reshaping by a reduced pulse propagator's encapsulation of the effects of a nonzero-duration resonant light pulse.

The wave-packet overlaps contributing to the WPI signal can be assembled from the appropriate one- and three-pulse kets. For instance, the SE and GSB overlaps in $P^{(s)}$ of eqn (6.9) are given, respectively, by

$$\langle\uparrow_A|\uparrow_B\downarrow_C\uparrow_D\rangle = -e^{-i\varphi_{DC}-i\varphi_{BA}}F_A F_B F_C F_D$$
$$\times\,\langle n_g|p_A^{(ge)}(t-t_A(\mathbf{r});\bar{\bar{\tau}})[-t_{DA}(\mathbf{r})]_{ee}\,p_D^{(eg)}(t-t_D(\mathbf{r});\tau)[t_{DC}(\mathbf{r})]_{gg}$$
$$\times\, p_C^{(ge)}(\tau+t_{DC}(\mathbf{r});\bar{\tau})[t_{CB}(\mathbf{r})]_{ee}\,p_B^{(eg)}(\bar{\tau}+t_{CB}(\mathbf{r});\bar{\bar{\tau}})[t_{BA}(\mathbf{r})]_{gg}|n_g\rangle\,, \quad (6.60)$$

and

$$\langle\uparrow_A\downarrow_B\uparrow_C|\uparrow_D\rangle = -e^{-i\varphi_{DC}-i\varphi_{BA}}F_A F_B F_C F_D$$
$$\times\,\langle n_g|p_A^{(ge)}(\bar{\tau}+t_{BA}(\mathbf{r});\bar{\bar{\tau}})[-t_{BA}(\mathbf{r})]_{ee}\,p_B^{(eg)}(\tau+t_{CB}(\mathbf{r});\bar{\tau})[-t_{CB}(\mathbf{r})]_{gg}$$
$$\times\, p_C^{(ge)}(t-t_C(\mathbf{r});\tau)[-t_{DC}(\mathbf{r})]_{ee}\,p_D^{(eg)}(t-t_D(\mathbf{r});\bar{\bar{\tau}})[t_{DA}(\mathbf{r})]_{gg}|n_g\rangle\,. \quad (6.61)$$

In fluorescence-detected WPI experiments, the four incident pulses often travel collinearly.[13] In this case (or if the experiment addresses a single molecule, whose location can be regarded as the origin), the interpulse delays become independent of the molecule's location: $t_{JI}(\mathbf{r}) = t_{JI}$. In addition, since the quadrilinear e-state population *after* all the pulses determines the amount of fluorescence, $t - t_K(\mathbf{r})$ can safely be replaced by infinity as the upper integration limit of the last-acting pulse propagators. Under these conditions, eqns (6.60) and (6.61) simplify to

$$\langle\uparrow_A|\uparrow_B\downarrow_C\uparrow_D\rangle = -e^{-i\varphi_{DC}-i\varphi_{BA}}F_A F_B F_C F_D$$
$$\times\,\langle n_g|p_A^{(ge)}(\infty;\bar{\bar{\tau}})[-t_{DA}]_{ee}\,p_D^{(eg)}(\infty;\tau)[t_{DC}]_{gg}$$
$$\times\, p_C^{(ge)}(\tau+t_{DC};\bar{\tau})[t_{CB}]_{ee}\,p_B^{(eg)}(\bar{\tau}+t_{CB};\bar{\bar{\tau}})[t_{BA}]_{gg}|n_g\rangle\,, \quad (6.62)$$

and

$$\langle\uparrow_A\downarrow_B\uparrow_C|\uparrow_D\rangle = -e^{-i\varphi_{DC}-i\varphi_{BA}}F_A F_B F_C F_D$$
$$\times\,\langle n_g|p_A^{(ge)}(\bar{\tau}+t_{BA};\bar{\bar{\tau}})[-t_{BA}]_{ee}\,p_B^{(eg)}(\tau+t_{CB};\bar{\tau})[-t_{CB}]_{gg}$$
$$\times\, p_C^{(ge)}(\infty;\tau)[-t_{DC}]_{ee}\,p_D^{(eg)}(\infty;\bar{\bar{\tau}})[t_{DA}]_{gg}|n_g\rangle\,, \quad (6.63)$$

[13] We will later consider a more general situation in which the four pulses are incident on the sample plane at small relative angles. This experimental geometry imposes a wave-vector-matching condition among the $\mathbf{n}_I$.

respectively. The corresponding simplified SE and GSB overlaps participating in $P^{(d)}$ are found, respectively, to be

$$\begin{aligned}\langle\uparrow_A\downarrow_C\uparrow_D\mid\uparrow_B\rangle &= -e^{i\varphi_{DC}-i\varphi_{BA}}F_AF_BF_CF_D \\ &\times\langle n_g|p_A^{(ge)}(\bar{\tau}+t_{CA};\bar{\bar{\tau}})[-t_{CA}]_{ee}\,p_C^{(eg)}(\tau+t_{DC};\bar{\tau})[-t_{DC}]_{gg} \\ &\times p_D^{(ge)}(\infty;\tau)[t_{DB}]_{ee}\,p_B^{(eg)}(\infty;\bar{\bar{\bar{\tau}}})[t_{BA}]_{gg}|n_g\rangle\,, \end{aligned} \tag{6.64}$$

and

$$\begin{aligned}\langle\uparrow_A\downarrow_B\uparrow_D\mid\uparrow_C\rangle &= -e^{i\varphi_{DC}-i\varphi_{BA}}F_AF_BF_CF_D \\ &\times\langle n_g|p_A^{(ge)}(\bar{\tau}+t_{BA};\bar{\bar{\tau}})[-t_{BA}]_{ee}\,p_B^{(eg)}(\tau+t_{DB};\bar{\tau})[-t_{DB}]_{gg} \\ &\times p_D^{(ge)}(\infty;\tau)[t_{DC}]_{ee}\,p_C^{(eg)}(\infty;\bar{\bar{\bar{\tau}}})[t_{CA}]_{gg}|n_g\rangle\,, \end{aligned} \tag{6.65}$$

while the simplified SE′ and GSB′ overlaps of $P^{(d)}$ take the respective forms

$$\begin{aligned}\langle\uparrow_D\mid\uparrow_C\downarrow_A\uparrow_B\rangle\} &= -e^{i\varphi_{DC}-i\varphi_{BA}}F_AF_BF_CF_D \\ &\times\langle n_g|[-t_{DA}]_{gg}\,p_D^{(ge)}(\infty;\bar{\bar{\bar{\tau}}})[t_{DB}]_{ee}\,p_B^{(eg)}(\infty;\tau)[t_{BA}]_{gg} \\ &\times p_A^{(ge)}(\tau+t_{BA};\bar{\tau})[-t_{CA}]_{ee}\,p_C^{(eg)}(\bar{\tau}-t_{CA};\bar{\bar{\tau}})[t_{CA}]_{gg}|n_g\rangle\,, \end{aligned} \tag{6.66}$$

and

$$\begin{aligned}\langle\uparrow_A\mid\uparrow_C\downarrow_D\uparrow_B\rangle &= -e^{i\varphi_{DC}-i\varphi_{BA}}F_AF_BF_CF_D \\ &\times\langle n_g|p_A^{(ge)}(\infty;\bar{\bar{\bar{\tau}}})[-t_{BA}]_{ee}\,p_B^{(eg)}(\infty;\tau)[-t_{DB}]_{gg} \\ &\times p_D^{(ge)}(\tau-t_{DB};\bar{\tau})[t_{DC}]_{ee}\,p_C^{(eg)}(\bar{\tau}+t_{DC};\bar{\bar{\tau}})[t_{CA}]_{gg}|n_g\rangle\,. \end{aligned} \tag{6.67}$$

The lack of dependence of any of the overlaps (6.62)–(6.67) upon t warrants specific notice: because both overlapped wave packets move under the same e-state nuclear Hamiltonian, the t-dependence of their separate unitary evolution cancels. This observation time enters the WPI overlaps only in the upper limit of the last-acting reduced pulse propagators, where it was replaced by infinity in these expressions. Moreover, as all location dependence vanishes from the contributing wave-packet overlaps under a collinear experimental geometry, 2D-WPI signals from a collection of noninteracting chromophores distributed throughout a spatially extended sample reduce to the sum of signals from the individual molecules.

> Find expressions analogous to eqns (6.62) through (6.67) for the remaining sum- and difference-phased wave-packet overlaps contributing to the quadrilinear e-state population.

We can gain an initial acquaintance with the dynamics controlling wave-packet overlaps by drawing plots of the $\langle q\rangle$ and $\langle p\rangle$ trajectories corresponding to the one- and three-pulse packets participating in the SE and GSB portions of $P^{(d)}$. Figure 6.6 shows the phase-space paths of $|\uparrow_A\downarrow_C\uparrow_D\rangle$ and $|\uparrow_B\rangle$ for the case of a one-dimensional

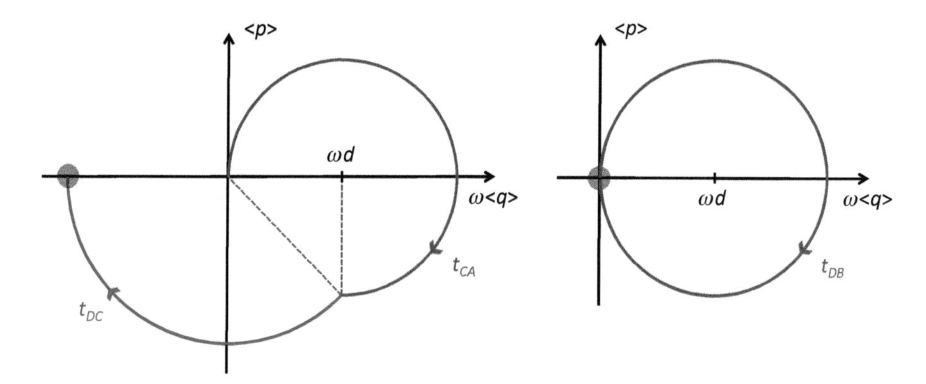

Fig. 6.6 Phase-space trajectories of three- and one-pulse wave packets whose overlap determines the difference-phased stimulated-emission contribution to the WPI signal in a one-dimensional displaced harmonic system. Blue paths are intervals of excited-state motion, red is a ground-state interval.

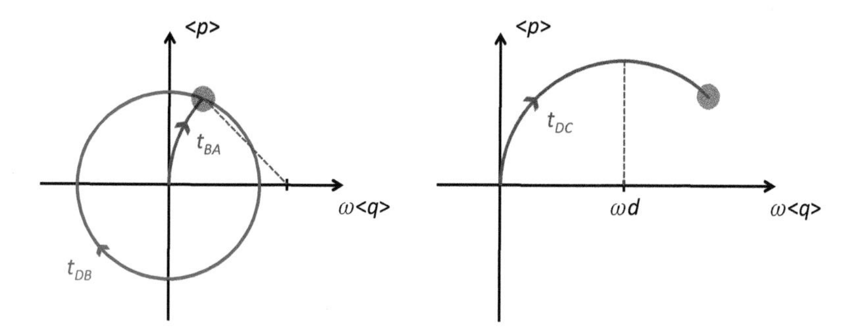

Fig. 6.7 Trajectories of the wave packets whose overlap specifies the ground-state-bleach portion of $P^{(d)}$ in the displaced-harmonic system.

system with $H_g = \frac{1}{2}p^2 + \frac{\omega^2}{2}q^2$ and $H_e = \frac{1}{2}p^2 + \frac{\omega^2}{2}(q-d)^2$, under the influence of vibrationally abrupt laser pulses. We have chosen $t_{BA} = \tau_v/8$, $t_{CB} = 5\tau_v/8$, and $t_{DC} = 3\tau_v/8$, where $\tau_v = 2\pi/\omega$ is the vibrational period. The overlap of these two wave packets specifies the SE overlap of $P^{(d)}$, but the *value* of this overlap as a function of the interpulse delays depends on the size of the e-state displacement d relative to the root-mean-squared width $\sqrt{\hbar/2\omega}$. Fig. 6.7 portrays the analogous phase-space trajectories for $|\uparrow_A\downarrow_B\uparrow_D\rangle$ and $|\uparrow_C\rangle$, whose overlap determines the GSB contribution to the difference-phased WPI signal.

Draw phase-space trajectories for the wave packets participating in the SE and GSB overlaps of $P^{(s)}$ for the same interpulse delay times as in Figs 6.6 and 6.7. Draw those for the overlapping wave packets of SE′ and GSB′ in $P^{(d)}$ with $t_{BA} = \tau_v/8$, $t_{CB} = -3\tau_v/16$, and $t_{DC} = 3\tau_v/8$.

We may proceed similarly to build the matrix elements of $\hat{\mathbf{m}}$ contributing to the

FWM signal. Among the contributions to $\mathbf{m}^{(s)}$, we focus on those elements in eqn (6.17) within which the pulses act in their nominal order (i.e., A before B before C) within each bra or ket, namely

$$\begin{aligned}\langle\uparrow_A\downarrow_B\uparrow_C|\hat{\mathbf{m}}|0\rangle = ie^{i\varphi_C - i\varphi_{BA}} m\mathbf{i}\, F_A F_B F_C \langle n_g|p_A^{(ge)}(\bar{\tau} + t_{BA}(\mathbf{r}); \bar{\bar{\tau}})[-t_{BA}(\mathbf{r})]_{ee} \\ \times p_B^{(eg)}(\tau + t_{CB}(\mathbf{r}); \bar{\tau})[-t_{CB}(\mathbf{r})]_{gg} p_C^{(ge)}(t - t_C(\mathbf{r}); \tau)[-t + t_C(\mathbf{r})]_{ee} \\ \times [t - t_A(\mathbf{r})]_{gg}|n_g\rangle\,, \end{aligned} \tag{6.68}$$

and

$$\begin{aligned}\langle\uparrow_A|\hat{\mathbf{m}}|\uparrow_B\downarrow_C\rangle = ie^{i\varphi_C - i\varphi_{BA}} m\mathbf{i}\, F_A F_B F_C \langle n_g|p_A^{(ge)}(t - t_A(\mathbf{r}); \bar{\bar{\tau}})[-t + t_A(\mathbf{r})]_{ee} \\ \times [t - t_C(\mathbf{r})]_{gg} p_C^{(ge)}(t - t_C(\mathbf{r}); \tau)[t_{CB}(\mathbf{r})]_{ee} p_B^{(eg)}(\tau + t_{CB}(\mathbf{r}); \bar{\tau})[t_{BA}(\mathbf{r})]_{gg}|n_g\rangle\,. \end{aligned} \tag{6.69}$$

The elements in $\mathbf{m}^{(d)}$ of eqn (6.18) for which the pulses act in their nominal order within each bra or ket are

$$\begin{aligned}\langle\uparrow_A\downarrow_C|\hat{\mathbf{m}}|\uparrow_B\rangle = -ie^{-i\varphi_C - i\varphi_{BA}} m\mathbf{i}\, F_A F_B F_C \langle n_g|p_A^{(ge)}(\tau + t_{CA}(\mathbf{r}); \bar{\tau})[-t_{CA}(\mathbf{r})]_{ee} \\ \times p_C^{(eg)}(t - t_C(\mathbf{r}); \tau)[-t + t_C(\mathbf{r})]_{gg} \\ \times [t - t_B(\mathbf{r})]_{ee} p_B^{(eg)}(t - t_B(\mathbf{r}); \bar{\bar{\tau}})[t_{BA}(\mathbf{r})]_{gg}|n_g\rangle\,, \end{aligned} \tag{6.70}$$

and

$$\begin{aligned}\langle\uparrow_A\downarrow_B|\hat{\mathbf{m}}|\uparrow_C\rangle = -ie^{-i\varphi_C - i\varphi_{BA}} m\mathbf{i}\, F_A F_B F_C \langle n_g|p_A^{(ge)}(\tau + t_{BA}(\mathbf{r}); \bar{\tau})[-t_{BA}(\mathbf{r})]_{ee} \\ \times p_B^{(eg)}(t - t_B(\mathbf{r}); \tau)[-t + t_B(\mathbf{r})]_{gg} \\ \times [t - t_C(\mathbf{r})]_{ee} p_C^{(eg)}(t - t_C(\mathbf{r}); \bar{\bar{\tau}})[t_{CA}(\mathbf{r})]_{gg}|n_g\rangle\,. \end{aligned} \tag{6.71}$$

The t-dependence of these trilinear dipole moments—in particular, their oscillation near the $e \leftarrow g$ transition frequency—is responsible for generating the FWM signal field. Their dependence on the molecular location $\mathbf{r}$ will be shown to give rise to the signal field's directionality.

> Write out formulas analogous to eqns (6.68)–(6.71) for the remaining matrix elements contributing to $\mathbf{m}^{(s)}$ and $\mathbf{m}^{(d)}$, those in which the pulses within a bra or ket act in other than their nominal order.

6.3.3 Wave-vector matching

A formula for the stimulated-emission overlap of $P^{(d)}$ more general than eqn (6.64) in being applicable to cases of noncollinear A, B, C, and D beams is

$$\begin{aligned}\langle\uparrow_A\downarrow_C\uparrow_D\,|\uparrow_B\rangle = -e^{i\varphi_{DC} - i\varphi_{BA}} F_A F_B F_C F_D \\ \times \langle n_g|p_A^{(ge)}(\bar{\tau} + t_{CA}(\mathbf{r}); \bar{\bar{\tau}})[-t_{CA}(\mathbf{r})]_{ee}\, p_C^{(eg)}(\tau + t_{DC}(\mathbf{r}); \bar{\tau})[-t_{DC}(\mathbf{r})]_{gg} \\ \times p_D^{(ge)}(\infty; \tau)[t_{DB}(\mathbf{r})]_{ee}\, p_B^{(eg)}(\infty; \bar{\bar{\bar{\tau}}})[t_{BA}(\mathbf{r})]_{gg}|n_g\rangle\,; \end{aligned} \tag{6.72}$$

the location dependence of the interpulse delays between noncollinearly propagating pulses has been restored. If all four propagation unit vectors are sufficiently close to the surface normal of a quasi-2D sample, then the interpulse delays $t_{JI}(\mathbf{r}) = t_{JI} + (\mathbf{n}_J - \mathbf{n}_I) \cdot \mathbf{r}/c$ will vary with molecular location at most by an amount $\sim \delta\theta_{JI} a/c$ that is negligible on the timescale of nuclear dynamics. Hence, the reduced pulse propagators can be evaluated at the sample origin (e.g., $p_C^{(eg)}(\tau + t_{DC}(\mathbf{r}); \bar{\tau}) \cong p_C^{(eg)}(\tau + t_{DC}; \bar{\tau})$), as can the ground-state free-evolution operators (e.g., $[t_{BA}(\mathbf{r})]_{gg} \cong [t_{BA}]_{gg}$). The only significant position dependence is attributable to the nonnegligibility of arrival-time variations on the electronic timescale and resides in the excited-state evolution operators (e.g., $[t_{DB}(\mathbf{r})]_{ee} \cong [t_{DB}]_{ee} \exp\{-i\Lambda(\mathbf{n}_D - \mathbf{n}_B) \cdot \mathbf{r}/c\}$).

In describing a fluorescence-detected WPI measurement on a collection of chromophores distributed with areal density α on a sample surface, we have, for example,

$$\begin{aligned} P_{SE}^{(d)} &= 2\alpha \mathrm{Re} \int d^2r \, \langle \uparrow_A \downarrow_C \uparrow_D \mid \uparrow_B \rangle_{\mathbf{r}} \\ &= 2\alpha \mathrm{Re} \left\{ \langle \uparrow_A \downarrow_C \uparrow_D \mid \uparrow_B \rangle_{\mathbf{0}} \int d^2r \, e^{i\Lambda(\mathbf{n}_C - \mathbf{n}_A)\cdot \mathbf{r}/c} e^{-i\Lambda(\mathbf{n}_D - \mathbf{n}_B)\cdot \mathbf{r}/c} \right\}. \end{aligned} \tag{6.73}$$

Under the stated conditions of near-normal incidence, both $\mathbf{n}_C - \mathbf{n}_A$ and $\mathbf{n}_D - \mathbf{n}_B$ are largely confined to the xy-plane. They must cancel each other in the exponent in order for $P_{SE}^{(d)}$ to take on nonnegligible values. We are led to the wave-vector-matching condition,

$$\mathbf{n}_D - \mathbf{n}_C - \mathbf{n}_B + \mathbf{n}_A = 0\,, \tag{6.74}$$

for the stimulated-emission term, and the same relation obtains in the other contributions to $P^{(d)}$ for an extended sample.

> Show that wave-vector matching imposes the requirement $\mathbf{n}_D - \mathbf{n}_C + \mathbf{n}_B - \mathbf{n}_A = 0$ on all the overlaps contributing to $P^{(s)}$ from a spatially extended sample.

Wave-vector matching imposes on the geometry of a FWM experiment the requirement that the signal field travel in the same direction as the LO (D) pulse. In the derivation of eqn (6.27), it was assumed that $\mathbf{n}_D = \mathbf{k}$ is perpendicular to the sample plane and $\mathbf{e}_D = \mathbf{i}$. We now further assume that the A, B, and C beams propagate at angles close to this surface normal. Under the approximations appropriate to this arrangement, similar to those brought to bear on the evaluation of the WPI signal, the spatial integral over the induced dipole of eqn (6.70) for instance, appearing in the FWM signal (6.27), becomes

$$\mathrm{Re} \int d^2r \, \langle \uparrow_A \downarrow_C | \hat{\mathbf{m}} | \uparrow_B \rangle_{\mathbf{r}} \cong \mathrm{Re} \left\{ \langle \uparrow_A \downarrow_C | \hat{\mathbf{m}} | \uparrow_B \rangle_{\mathbf{0}} \int d^2r \, e^{i\Lambda(\mathbf{n}_C - \mathbf{n}_A)\cdot \mathbf{r}/c} e^{i\Lambda \mathbf{n}_B \cdot \mathbf{r}/c} \right\}. \tag{6.75}$$

For this integral over the sample area to take a nonnegligible value requires $(\mathbf{n}_C - \mathbf{n}_A + \mathbf{n}_B) \cdot \mathbf{i} = (\mathbf{n}_C - \mathbf{n}_A + \mathbf{n}_B) \cdot \mathbf{j} = 0$, or

$$\mathbf{n}_C - \mathbf{n}_A + \mathbf{n}_B = \mathbf{k}\,, \tag{6.76}$$

which, in the present arrangement, is the same as eqn (6.74).

1. Show that eqn (6.76) applies also to the portion of $\mathbf{m}^{(d)}$ arising from eqn (6.71).
2. Find the wave-vector-matching condition for $\mathbf{m}^{(s)}$.

When the wave-vector-matching condition of eqn (6.76) is met, the induced dipole (eqn (6.75)) makes the following contribution to the FWM signal:

$$\Delta\mathcal{U}_{SE}^{(d)} = -\mathcal{N} E_D \Omega \,\mathrm{Re} \int_{-\infty}^{\infty} dt\, f_D(t-t_D) \sin\left[\Omega(t-t_D)+\varphi_D\right] \langle \uparrow_A \downarrow_C |\mathbf{i}\cdot\hat{\mathbf{m}}| \uparrow_B \rangle\,, \quad (6.77)$$

where $\mathcal{N} = \alpha \int d^2 r$ counts the number of chromophores in the laser spot, and the trilinear matrix element is that for a molecule located at the origin. Inserting eqn (6.70) for this element and making a rotating-wave approximation then gives

$$\Delta\mathcal{U}_{SE}^{(d)} = \mathcal{N}\hbar\Omega F_A F_B F_C F_D \,\mathrm{Re}\{e^{i\varphi_{DC}-i\varphi_{BA}} \langle n_g | p_A^{(ge)}(\tau + t_{CA}; \bar{\tau})[-t_{CA}]_{ee}$$
$$\times\, p_C^{(eg)}(\bar{t} + t_{DC}; \tau)[-t_{DC}]_{gg}\, p_D^{(ge)}(\infty; \bar{t})[t_{DB}]_{ee}\, p_B^{(eg)}(\bar{t} + t_{DB}; \bar{\bar{\tau}})[t_{BA}]_{gg} | n_g\rangle\}\,. \quad (6.78)$$

The form of this expression justifies its SE label. But notice that the integration variable of the D-pulse propagator appears in the upper limits of *both* the C and B propagators; this reflects the fact that the local oscillator participates only by interfering with the radiated trilinear signal field in a FWM experiment. In particular, the LO can never act before the B pulse, as it can in the corresponding WPI-signal contribution given by the real part of the overlap in eqn (6.64).

1. Plot the region of $\{t_{BA}, t_{DC}, t_{CB}\}$ space within which this distinction between $\Delta\mathcal{U}_{SE}^{(d)}$ and $P_{SE}^{(d)}$ may become significant.
2. Find an expression for the GSB contribution to $\Delta\mathcal{U}^{(d)}$ and determine whether and when it may differ in information content from $P_{GSB}^{(d)}$, which is determined by the quadrilinear overlap given in eqn (6.65).

6.3.4 Intrapulse-pair delay dependence

It is a common practice in 2D-WPI or FWM spectroscopy to Fourier transform P or $\Delta\mathcal{U}$, respectively, with respect to the intrapulse-pair delays t_{BA} and t_{DC}.[14] t_{CB} is regarded as the pump-probe delay or "waiting time," to be held constant while carrying out the two-dimensional Fourier transformation. Such work-ups yield two-dimensional, complex-valued frequency-versus-frequency interferograms in which the dynamics of "diagonal" ($\omega_{BA} \approx \omega_{DC}$) and "off-diagonal" ($\omega_{BA} \neq \omega_{DC}$) features can be studied as a function of the delay between the pump and probe pulse-pairs. Instead of eqn (6.64) for the overlap determining the stimulated-emission contribution to $P^{(d)}$, for example, one would make use of an equivalent formula,

[14]Fourier transformation involves integration over the full range of a time variable from negative to positive infinity. In our accounting, with nonnegative intrapulse-pair delays only, "negative t_{BA}" would be replaced by positive t_{BA} with φ_{BA} changed in sign, and so forth, as described in Sec. 6.2.1.

$$\langle \uparrow_A \downarrow_C \uparrow_D \mid \uparrow_B \rangle = -e^{i\varphi_{DC} - i\varphi_{BA}} F_A F_B F_C F_D$$
$$\times \langle n_g | p_A^{(ge)}(\bar{\tau} + t_{CB} + t_{BA}; \bar{\bar{\tau}})[-t_{BA}]_{ee}[-t_{CB}]_{ee} p_C^{(eg)}(\tau + t_{DC}; \bar{\tau})[-t_{DC}]_{gg}$$
$$\times p_D^{(ge)}(\infty; \tau)[t_{DC}]_{ee}[t_{CB}]_{ee} p_B^{(eg)}(\infty; \bar{\bar{\bar{\tau}}})[t_{BA}]_{gg} | n_g \rangle , \quad (6.79)$$

which makes the delays t_{BA}, t_{DC}, and t_{CB} explicit.

There is no *a priori* requirement that either experimental data or simulations of 2D spectroscopy signals be organized in frequency-versus-frequency plots. These data could equally well be displayed as t_{BA}-versus-t_{DC} interferograms with specified values of t_{CB} (or as mixed time-versus-frequency interferograms). It is advantageous in some respects to plot 2D data as intrapulse-pair time-versus-time plots: As noted in Section 6.2.1, some of the overlaps contributing to a whoopee signal vanish entirely except in narrow regions of $\{t_{BA}, t_{DC}, t_{CB}\}$. This property directly simplifies the interpretation of time-versus-time interferograms, but manifests itself less directly in frequency-versus-frequency plots.

6.4 Example signals

6.4.1 Short-pulse limit

It is helpful to examine 2D-WPI signals from model systems consisting of a few vibrational modes whose equilibrium positions are displaced by certain distances in the electronic excited state. In order to simplify these illustrative calculations, we specialize to a short-pulse limit in which the laser pulses are electronically resonant and their durations $\sim\sigma$ are sufficiently brief that nuclear motion inside a pulse propagator, the effects of temporal pulse overlap, and the nonzero turn-on time of the various wave-packet overlaps contributing to the WPI signal can all be neglected. The basic expressions of eqns (6.9) and (6.10) for the 2D-WPI signal simplify markedly in this limit—due the facts that pulse B cannot act before pulse A, pulse D cannot act before pulse C, and pulse D certainly cannot act before pulse A—becoming

$$P^{(s)} = 2\mathrm{Re}\{\underbrace{\langle \uparrow_A \mid \uparrow_B \downarrow_C \uparrow_D \rangle}_{\mathrm{SE}} + \underbrace{\langle \uparrow_A \downarrow_B \uparrow_C \mid \uparrow_D \rangle}_{\mathrm{GSB}}\} , \quad (6.80)$$

and

$$P^{(d)} = 2\mathrm{Re}\{\underbrace{\langle \uparrow_A \downarrow_C \uparrow_D \mid \uparrow_B \rangle}_{\mathrm{SE}} + \underbrace{\langle \uparrow_A \downarrow_B \uparrow_D \mid \uparrow_C \rangle}_{\mathrm{GSB}}$$
$$+ \underbrace{\langle \uparrow_D \mid \uparrow_C \downarrow_A \uparrow_B \rangle}_{\mathrm{SE}'} + \underbrace{\langle \uparrow_A \mid \uparrow_C \downarrow_D \uparrow_B \rangle}_{\mathrm{GSB}'}\} , \quad (6.81)$$

respectively. The SE′ and GSB′ terms in $P^{(d)}$ have been kept because, in a complete data set, it is possible for t_C to precede t_A or t_D to precede t_B (see Section 6.2.1). In effect, with regard to $P^{(d)}$ specifically, the temporal regions of possibly nonvanishing overlap shown in Figs 6.1 and 6.2 become simpler as σ goes to zero, while those in Figs 6.3 and 6.4 disappear entirely.

The assumption of arbitrarily short pulses also simplifies the quadrilinear overlaps themselves. An unnested reduced pulse propagator, for example, becomes

$$p_I^{(eg)}(\infty;\tau) = \int_{-\infty}^{\infty} \frac{d\tau}{\sigma} f_I(\tau) \equiv \alpha_I\,. \tag{6.82}$$

A combination of two nested pulse propagators simplifies to

$$\begin{aligned} p_J^{(ge)}(\infty;\tau)\ldots p_I^{(eg)}(\tau+t_{JI};\bar{\tau}) &= \int_{-\infty}^{\infty} \frac{d\tau}{\sigma} f_J(\tau)\cdots \int_{-\infty}^{\tau+t_{JI}} \frac{d\bar{\tau}}{\sigma} f_I(\bar{\tau}) \\ &\equiv \alpha_J \alpha_I \Theta(t_{JI})\ldots\,; \end{aligned} \tag{6.83}$$

because the reduced propagators are dimensionless real numbers in the short-pulse limit, they commute with the intervening time-evolution operator (signified by an ellipsis).

From eqn (6.62) for the sum-phased SE overlap, we find

$$\begin{aligned} \langle \uparrow_A \mid \uparrow_B \downarrow_C \uparrow_D \rangle &= -e^{-i\varphi_{DC}-i\varphi_{BA}} \alpha_A\alpha_B\alpha_C\alpha_D\, F_A F_B F_C F_D\, \Theta(t_{CB}) \\ &\quad \times \langle n_g | [-t_{DA}]_{ee} [t_{DC}]_{gg} [t_{CB}]_{ee} [t_{BA}]_{gg} | n_g \rangle\,, \end{aligned} \tag{6.84}$$

and from eqn (6.63) for the sum-phased GSB,

$$\begin{aligned} \langle \uparrow_A \downarrow_B \uparrow_C \mid \uparrow_D \rangle &= -e^{-i\varphi_{DC}-i\varphi_{BA}} \alpha_A\alpha_B\alpha_C\alpha_D\, F_A F_B F_C F_D\, \Theta(t_{CB}) \\ &\quad \times \langle n_g | [-t_{BA}]_{ee} [-t_{CB}]_{gg} [-t_{DC}]_{ee} [t_{DA}]_{gg} | n_g \rangle\,. \end{aligned} \tag{6.85}$$

Next, we turn to the four wave-packet overlaps contributing to $P^{(d)}$ in eqn (6.81). The one responsible for the short-pulse, difference-phased SE signal is found, from eqn (6.64) or (6.79), to be

$$\begin{aligned} \langle \uparrow_A \downarrow_C \uparrow_D \mid \uparrow_B \rangle &= -e^{i\varphi_{DC}-i\varphi_{BA}} \alpha_A\alpha_B\alpha_C\alpha_D\, F_A F_B F_C F_D\, \Theta(t_{CA}) \\ &\quad \times \langle n_g | [-t_{CA}]_{ee} [-t_{DC}]_{gg} [t_{DB}]_{ee} [t_{BA}]_{gg} | n_g \rangle\,. \end{aligned} \tag{6.86}$$

From eqn (6.65), the overlap giving rise to the difference-phased GSB signal becomes

$$\begin{aligned} \langle \uparrow_A \downarrow_B \uparrow_D \mid \uparrow_C \rangle &= -e^{i\varphi_{DC}-i\varphi_{BA}} \alpha_A\alpha_B\alpha_C\alpha_D\, F_A F_B F_C F_D\, \Theta(t_{DB}) \\ &\quad \times \langle n_g | [-t_{BA}]_{ee} [-t_{DB}]_{gg} [t_{DC}]_{ee} [t_{CA}]_{gg} | n_g \rangle\,. \end{aligned} \tag{6.87}$$

Similar formulas are easily obtained for the overlaps determining the remaining, less important contributions to $P^{(d)}$. These are the SE$'$ overlap,

$$\begin{aligned} \langle \uparrow_D \mid \uparrow_C \downarrow_A \uparrow_B \rangle &= -e^{i\varphi_{DC}-i\varphi_{BA}} \alpha_A\alpha_B\alpha_C\alpha_D\, F_A F_B F_C F_D\, \Theta(t_{AC}) \\ &\quad \times \langle n_g | [-t_{DA}]_{gg} [t_{DB}]_{ee} [t_{BA}]_{gg} [t_{AC}]_{ee} [t_{CA}]_{gg} | n_g \rangle\,, \end{aligned} \tag{6.88}$$

and the GSB$'$ overlap,

$$\begin{aligned} \langle \uparrow_A \mid \uparrow_C \downarrow_D \uparrow_B \rangle &= -e^{i\varphi_{DC}-i\varphi_{BA}} \alpha_A\alpha_B\alpha_C\alpha_D\, F_A F_B F_C F_D\, \Theta(t_{BD}) \\ &\quad \times \langle n_g | [-t_{BA}]_{ee} [t_{BD}]_{gg} [t_{DC}]_{ee} [t_{CA}]_{gg} | n_g \rangle\,. \end{aligned} \tag{6.89}$$

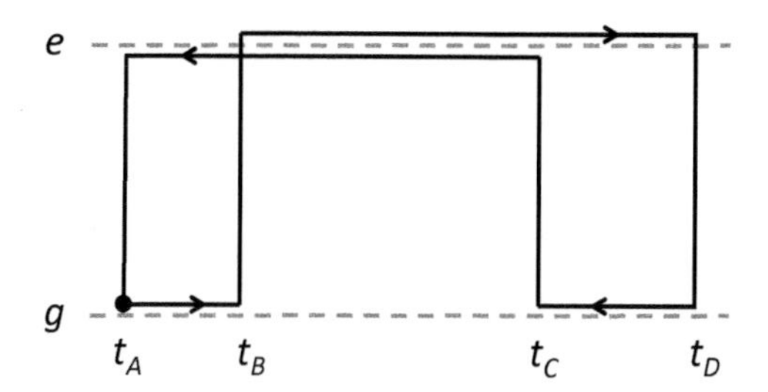

Fig. 6.8 Graphical depiction of the sequence of ground- and excited-state wave-packet evolutions within $\langle n_g|[-t_{CA}]_{ee}[-t_{DC}]_{gg}[t_{DB}]_{ee}[t_{BA}]_{gg}|n_g\rangle$. Forward motion starts in the g-state at time t_A and switches to the e-state at t_B. The wave packet returns to g at time t_D, where it undergoes backward evolution until t_C; backward motion then continues in the e-state until t_A, when the overlap between the propagated wave packet and $|n_g\rangle$ is evaluated.

6.4.2 Cumulant expansion

In this subsection, we develop a practical approximate approach to calculating the wave-packet overlaps determining 2D-WPI signals. This *second-cumulant approximation* is illustrated with the short-pulse limit of the overlap $\langle \uparrow_A \downarrow_C \uparrow_D \mid \uparrow_B \rangle$, given by eqn (6.86), which generates the stimulated-emission contribution to the difference-phased WPI signal (eqn (6.81)). From the time-circuit diagram[15] in Fig. 6.8, we see that the quantity $\langle n_g|[-t_{CA}]_{ee}[-t_{DC}]_{gg}[t_{DB}]_{ee}[t_{BA}]_{gg}|n_g\rangle$ to which this overlap is proportional comprises equal and opposite net amounts of g-state $(t_{BA} - t_{DC})$ and e-state $(t_{DC} - t_{BA})$ evolution.

This feature allows us to break the overall ket-to-bra time evolution into a sequence of equal- and opposite-duration episodes of g- and e-state motion. For example, we can write

$$\begin{aligned} &\langle n_g|[-t_{CA}]_{ee}[-t_{DC}]_{gg}[t_{DB}]_{ee}[t_{BA}]_{gg}|n_g\rangle \\ &\quad = \langle n_g|\left([-t_{CA}]_{ee}[t_{CA}]_{gg}\right)\left([-t_{DA}]_{gg}[t_{DA}]_{ee}\right)\left([-t_{BA}]_{ee}[t_{BA}]_{gg}\right)|n_g\rangle \\ &\quad = \langle n_g|d^\dagger(t_{CA})d(t_{DA})d^\dagger(t_{BA})|n_g\rangle \,, \end{aligned} \tag{6.90}$$

using the notation,

$$d(t) \equiv [-t]_{gg}[t]_{ee} \,. \tag{6.91}$$

The alternative time-path corresponding to the last member of eqn (6.90) is sketched in Fig. 6.9. We wish to calculate this and other similar matrix elements rigorously through second order, and approximately at higher orders, with respect to

$$v = H_e - H_g - \langle n_g|\left(H_e - H_g\right)|n_g\rangle \,; \tag{6.92}$$

this putatively small quantity is the coordinate-dependent electronic transition energy less its average value in the initial vibrational eigenstate. To accomplish this goal, we rewrite eqn (6.91) in the form

[15] For an early example of their use, see J. Almy, K. Kizer, R. Zadoyan, and V. A. Apkarian, "Resonant Raman, hot, and cold luminescence of iodine in rare gas matrixes," J. Phys. Chem A **104**, 3508–3520 (2000).

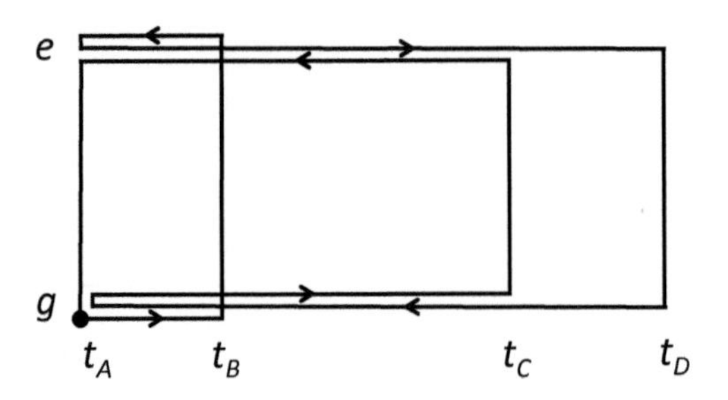

Fig. 6.9 Alternative time-circuit diagram representing the stimulated-emission overlap $\langle n_g|[-t_{CA}]_{ee}[-t_{DC}]_{gg}[t_{DB}]_{ee}[t_{BA}]_{gg}|n_g\rangle$ as a concatenation of equal-duration forward and backward wave-packet propagations according to last line of eqn (6.90).

$$\begin{aligned} d(t) &= e^{-i\Lambda t} e^{iH_g t/\hbar} e^{-iH_e t/\hbar} \\ &= e^{-i\bar{\Lambda} t} e^{iH_g t/\hbar} e^{-i(H_g+v)t/\hbar}, \end{aligned} \tag{6.93}$$

where $\bar{\Lambda} = \Lambda + \frac{1}{\hbar}\langle n_g|\,(H_e - H_g)\,|n_g\rangle$. A straightforward application of time-dependent perturbation theory leads to

$$d(t) \cong e^{-i\bar{\Lambda} t}\left\{1 - \frac{i}{\hbar}\int_0^t d\tau\, v(\tau) - \frac{1}{\hbar^2}\int_0^t d\tau \int_0^\tau d\bar{\tau}\, v(\tau)v(\bar{\tau})\right\}, \tag{6.94}$$

and

$$d^\dagger(t) \cong e^{i\bar{\Lambda} t}\left\{1 + \frac{i}{\hbar}\int_0^t d\tau\, v(\tau) - \frac{1}{\hbar^2}\int_0^t d\tau \int_0^\tau d\bar{\tau}\, v(\bar{\tau})v(\tau)\right\}, \tag{6.95}$$

with $v(t) \equiv e^{iH_g t/\hbar} v e^{-iH_g t/\hbar}$. Through second order in v, eqn (6.90) then becomes

$$\begin{aligned} &\langle n_g|[-t_{CA}]_{ee}[-t_{DC}]_{gg}[t_{DB}]_{ee}[t_{BA}]_{gg}|n_g\rangle \cong e^{-i\bar{\Lambda} t_{DC} + i\bar{\Lambda} t_{BA}} \\ &\times \left\{1 - g(-t_{CA}) - g(t_{DA}) - g(-t_{BA}) + h(t_{CA}, t_{DA}) - h(t_{CA}, t_{BA}) + h(t_{DA}, t_{BA})\right\}. \end{aligned} \tag{6.96}$$

This last equation uses

$$g(t) \equiv \frac{1}{\hbar^2}\int_0^t d\tau \int_0^\tau d\bar{\tau}\, \langle n_g|v(\tau)v(\bar{\tau})|n_g\rangle, \tag{6.97}$$

and

$$h(t_1, t_2) \equiv \frac{1}{\hbar^2}\int_0^{t_1} d\tau \int_0^{t_2} d\bar{\tau}\, \langle n_g|v(\tau)v(\bar{\tau})|n_g\rangle, \tag{6.98}$$

along with $\langle n_g|v(\tau)|n_g\rangle = 0$ and $g^*(t) = g(-t)$. The function h defined in eqn (6.98) can be re-expressed as

$$h(t_1, t_2) = g(t_1) + g(-t_2) - g(t_1 - t_2). \tag{6.99}$$

Verify eqns (6.96) and (6.99), proving in particular that $g^*(t) = g(-t)$.

Upon substitution in eqn (6.96) and re-exponentiation, we obtain a cumulant expansion for the SE overlap of $P^{(d)}$:

$$\begin{aligned}\langle \uparrow_A \downarrow_C \uparrow_D \mid \uparrow_B \rangle &\cong -e^{i\varphi_{DC} - i\varphi_{BA} - i\bar{\Lambda} t_{DC} + i\bar{\Lambda} t_{BA}} \alpha_A \alpha_B \alpha_C \alpha_D \, F_A F_B F_C F_D \, \Theta(t_{CA}) \\ &\times \exp\{g(t_{CB}) + g(-t_{DA}) - g(t_{DB}) - g(-t_{CA}) - g(-t_{BA}) - g(-t_{DC})\} . \quad (6.100)\end{aligned}$$

Re-exponentiation gives rise to contributions to the overlap of fourth- and higher-order in v which are lacking in eqn (6.96); in the cumulant approximation, the argument of the exponent, rather than the overlap itself, is given correctly through second order in v. But the validity of either approximation rests on the smallness of the electronic difference potential; neither eqn (6.96) nor eqn (6.100) is guaranteed to be correct in general.[16]

Similar evaluations could be carried through for the other three overlaps of eqns (6.86)–(6.89). But it is easier to obtain them from eqn (6.100) by permuting indices. Switching the B and C pulse labels (but keeping the correct step function) gives a cumulant expansion for the difference-phased GSB overlap:

$$\begin{aligned}\langle \uparrow_A \downarrow_B \uparrow_D \mid \uparrow_C \rangle &\cong -e^{i\varphi_{DC} - i\varphi_{BA} - i\bar{\Lambda} t_{DC} + i\bar{\Lambda} t_{BA}} \alpha_A \alpha_B \alpha_C \alpha_D \, F_A F_B F_C F_D \, \Theta(t_{DB}) \\ &\times \exp\{g(-t_{CB}) + g(-t_{DA}) - g(t_{DC}) - g(-t_{BA}) - g(-t_{CA}) - g(-t_{DB})\} . \quad (6.101)\end{aligned}$$

The exchanges $A \leftrightarrow C$ and $B \leftrightarrow D$, plus complex conjugation, convert eqn (6.100) into the SE$'$ overlap of $P^{(d)}$:

$$\begin{aligned}\langle \uparrow_D \mid \uparrow_C \downarrow_A \uparrow_B \rangle &\cong -e^{i\varphi_{DC} - i\varphi_{BA} - i\bar{\Lambda} t_{DC} + i\bar{\Lambda} t_{BA}} \alpha_A \alpha_B \alpha_C \alpha_D \, F_A F_B F_C F_D \, \Theta(t_{AC}) \\ &\times \exp\{g(t_{DA}) + g(-t_{CB}) - g(t_{DB}) - g(t_{AC}) - g(t_{DC}) - g(t_{BA})\} . \quad (6.102)\end{aligned}$$

The same exchange of pulse labels and complex conjugation turn eqn (6.101) into the short-pulse, difference-phased GSB$'$ overlap:

$$\begin{aligned}\langle \uparrow_A \mid \uparrow_C \downarrow_D \uparrow_B \rangle &\cong -e^{i\varphi_{DC} - i\varphi_{BA} - i\bar{\Lambda} t_{DC} + i\bar{\Lambda} t_{BA}} \alpha_A \alpha_B \alpha_C \alpha_D \, F_A F_B F_C F_D \, \Theta(t_{BD}) \\ &\times \exp\{g(-t_{DA}) + g(-t_{CB}) - g(-t_{BA}) - g(t_{DC}) - g(-t_{CA}) - g(t_{BD})\} . \quad (6.103)\end{aligned}$$

Derive second-cumulant approximations to the short-pulse, sum-phased SE and GSB overlaps given in eqns (6.84) and (6.85), respectively.

6.4.3 SE interferogram for a 1D displaced-oscillator system

Although the WPI signal $P^{(d)}$ of eqn (6.81) is a sum of four terms, we will examine just the SE contribution for ease of interpretation. We consider a system having a single molecular vibration, with harmonic potentials of frequency ω whose equilibrium positions in the ground and excited electronic states are shifted by $d = 1.2\sqrt{2\hbar/\omega}$ (see page 81). For this system, the second-cumulant "approximation" to

[16] A detailed treatment of nonlinear optical response functions using cumulant-expansion methods is given in Chap. 8 of S. Mukamel, *Principles of Nonlinear Optical Spectroscopy* (Oxford University Press, New York, 1999).

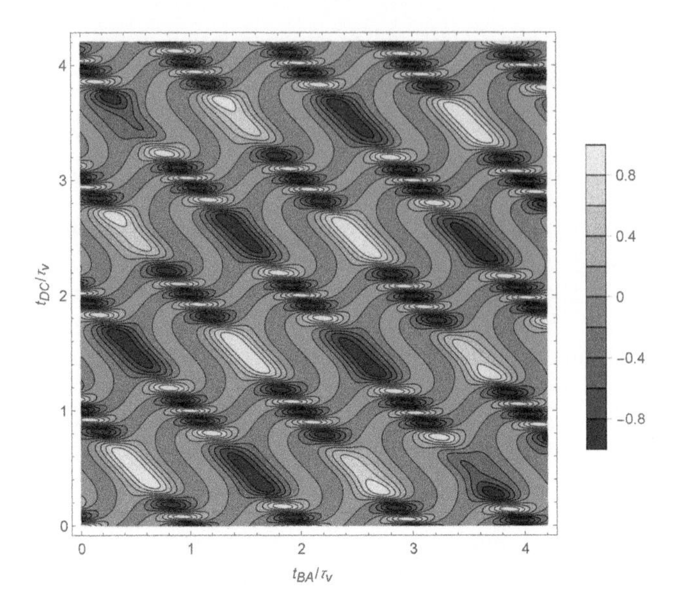

Fig. 6.10 Real part of overlap determining SE contribution to $P^{(d)}$ WPI signal for one-dimensional displaced harmonic system with dimensionless displacement $\delta = 1.2$, calculated in the short-pulse limit at $t_{CB} = 2.5\,\tau_v$.

the wave-packet overlaps is actually exact.[17] Figure 6.10 displays the real part of $e^{i\bar{\Lambda}(t_{DC}-t_{BA})}\langle 0_g|[-t_{CA}]_{ee}[-t_{DC}]_{gg}[t_{DB}]_{ee}[t_{BA}]_{gg}|0_g\rangle$ as a function of t_{BA} and t_{DC} for $t_{CB} = 2.5\,\tau_v$, where $\tau_v = 2\pi/\omega$ is the oscillator's period (see eqns (6.90), (6.100), and the exercise below; notice the aliasing at the vertical electronic transition frequency $\bar{\Lambda} = \Lambda + \omega\delta^2$). A similar interferogram is plotted in Fig. 6.10 for the waiting time $t_{CB} = -\tau_v$. Here, the absence of plotted signal for small t_{BA} can be understood by considering the $t_{CB} = -\tau_v$ slice through the SE delay volume (viz., the mirror image through $t_{DC} = t_{BA}$ of the volume shown in Fig. 6.1, in the limit $\sigma = 0$).

> Use an expression for the difference potential $v = H_e - H_g$ in terms of the creation ($a_g^\dagger$) and annihilation (a_g) operators for vibrational quanta in the ground-state oscillator to show that $g(t) = \delta^2(1 - i\omega t - e^{-i\omega t})$, where $\delta = d\sqrt{\omega/2\hbar}$, for the 1D displaced harmonic system.

If the displacement d is larger than the zero-point width of the wave packet, $\sqrt{\hbar/2\omega}$, as it is in the present example, then an overlap can be non-negligible only if the phase-points $(\langle q(t)\rangle, \langle p(t)\rangle)$ of the three-pulse and one-pulse wave packets are in near agreement. For the SE overlap of $P^{(d)}$, consideration of phase-space plots like those in Fig. 6.6 shows that there are two basic ways in which this phase-point coincidence can occur. Either (1) the phase-space locations after t_{CA} and t_{DB} of e-state evolution are

[17] They can be calculated directly by describing the successive episodes of ground- and excited-state motion using *Glauber coherent states.* See Complement G_V of C. Cohen-Tannoudji, B. Liu, and F. Laloë, *Quantum Mechanics* (Hermann, Paris, 1977).

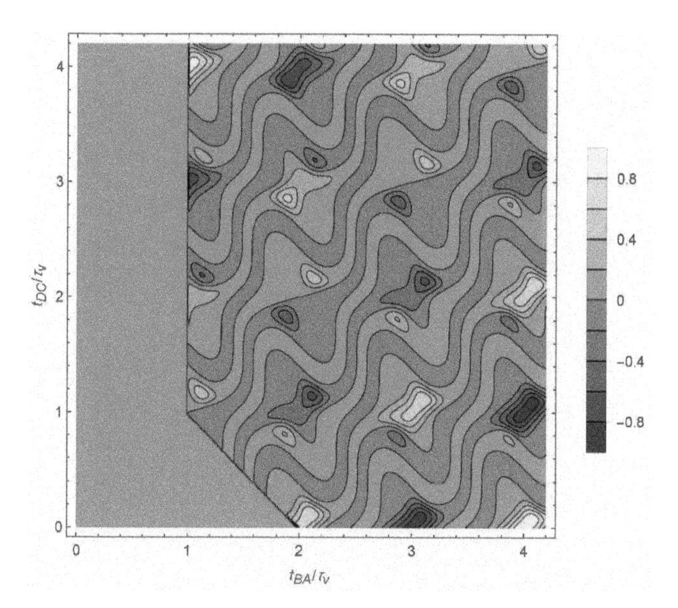

Fig. 6.11 Imaginary part of overlap determining SE contribution to $P^{(d)}$ for the one-dimensional harmonic system, in the short-pulse limit at $t_{CB} = -\tau_v$.

very similar, and a complete period or periods of subsequent g-state motion during t_{DC} returns the former to the same phase-point, or else (2) the two phase-points after t_{CA} and t_{DB} of motion in the e-state have the same coordinate value and opposite momenta, and t_{DC} of subsequent g-state evolution carries the former into the vicinity of the latter.

In either instance, full periods of oscillatory motion can be added to any of the intervals without affecting the coincidence in phase space. Thus the condition 1 for non-negligible SE overlap can be written,

$$\{t_{CA}/\tau_v\} = \{t_{DB}/\tau_v\} \text{ and } \{t_{DC}/\tau_v\} = 0\,; \tag{6.104}$$

where $\tau_v = 2\pi/\omega$ is the vibrational period and $\{x\}$ is the fractional excess of x, meaning the difference between x and the largest integer less than or equal to it (note that $0 \leq \{x\} < 1$). In the special circumstance of $\{t_{CA}/\tau_v\} = \{t_{DB}/\tau_v\} = 0$, corresponding to complete periods of e-state motion, the first phase-point becomes stationary upon transfer to the g-state and the value of t_{DC} is immaterial.

In case 2, the fractional excesses for the initial episodes of excited-state evolution differ from each other and add to unity:

$$\{t_{CA}/\tau_v\} + \{t_{DB}/\tau_v\} = 1\,. \tag{6.105}$$

There are two different versions of this case. In case 2a (2b), $\{t_{CA}/\tau_v\}$ is less than (greater than) $\{t_{DB}/\tau_v\}$. In both versions, t_{DC} of g-state motion must bring the first phase-point into coincidence with the second in order to produce non-negligible overlap, as shown in Fig. 6.12. In case 2a, the angle from $(\omega d,\, 0)$ between the phase-

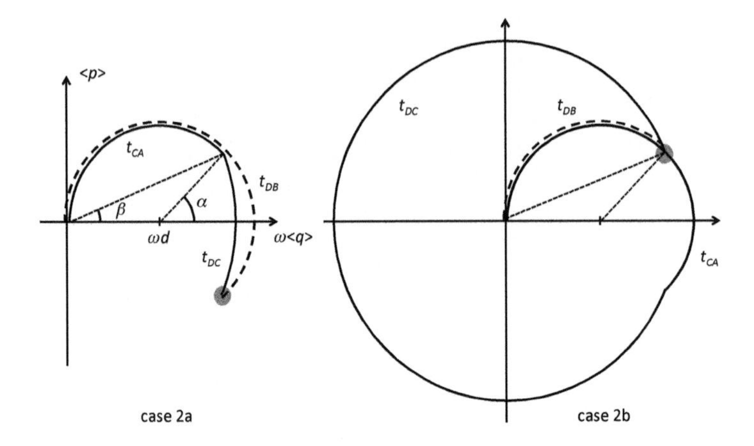

Fig. 6.12 Two types of case-2 phase-space trajectories leading to non-negligible stimulated-emission contribution to $P^{(d)}$. In both plots, trajectory for three-pulse wave packet is given by solid curve, while that for one-pulse packet is dashed.

point of the three-pulse wave packet after t_{CA} and the ωq-axis—α in the figure—is twice the angle from (0, 0) between this point and the ωq-axis—β in the figure. A second condition for case-2a coincidence therefore becomes

$$2\{t_{DC}/\tau_v\} + \{t_{CA}/\tau_v\} = \{t_{DB}/\tau_v\}. \tag{6.106}$$

The corresponding condition for case-2b coincidence is readily found to be

$$2\{t_{DC}/\tau_v\} + \{t_{CA}/\tau_v\} = 2 + \{t_{DB}/\tau_v\}. \tag{6.107}$$

Carry through the arguments leading to eqns (6.105) through (6.107).

Let's apply these arguments to the interferogram of Fig. 6.10. Near the anti-diagonally elongated peaks at integer periods in both t_{BA} and t_{DC}, we have $t_{BA} = (m - \chi)\tau_v$ and $t_{DC} = (n + \chi)\tau_v$, where m and n are integers and $|\chi| \ll 1$, whence

$$\left\{\frac{t_{DC}}{\tau_v}\right\} = \begin{cases} \chi & \text{if } \chi < 0 \\ 0 & \text{if } \chi = 0 \\ 1 + \chi & \text{if } \chi < 0 \end{cases}. \tag{6.108}$$

Since $t_{CB} = 2.5\,\tau_v$ for this plot,

$$\{t_{CA}/\tau_v\} = 0.5 - \chi \quad \text{and} \quad \{t_{DB}/\tau_v\} = 0.5 + \chi \tag{6.109}$$

in the vicinity of each of these peaks. In consequence, $\{t_{CA}/\tau_v\} + \{t_{DB}/\tau_v\} = 1$ as in eqn (6.105), and

$$2\left\{\frac{t_{DC}}{\tau_v}\right\} + \left\{\frac{t_{CA}}{\tau_v}\right\} - \left\{\frac{t_{DB}}{\tau_v}\right\} = \begin{cases} 0 & \chi > 0 \\ 0 & \chi = 0 \\ 2 & \chi < 0 \end{cases}, \tag{6.110}$$

corresponding to case 2a when $\chi > 0$, case 1 when $\chi = 0$, and case 2b when $\chi < 0$.

Along the anti-diagonally-shaped peaks at odd half-integer multiples of τ_v for both t_{BA} and t_{DC} in Fig. 6.10, we have $\{t_{BA}/\tau_v\} = 0.5 - \chi$ and $\{t_{DC}/\tau_v\} = 0.5 + \chi$. In addition,

$$\left\{\frac{t_{DB}}{\tau_v}\right\} = \begin{cases} \chi & \chi > 0 \\ 0 & \chi = 0 \\ 1+\chi & \chi < 0 \end{cases} \tag{6.111}$$

and

$$\left\{\frac{t_{CA}}{\tau_v}\right\} = \begin{cases} 1-\chi & \chi > 0 \\ 0 & \chi = 0 \\ -\chi & \chi < 0 \end{cases} . \tag{6.112}$$

Hence,

$$2\left\{\frac{t_{DC}}{\tau_v}\right\} + \left\{\frac{t_{CA}}{\tau_v}\right\} - \left\{\frac{t_{DB}}{\tau_v}\right\} = \begin{cases} 2 & \chi > 0 \\ 1 & \chi = 0 \\ 0 & \chi < 0 \end{cases} , \tag{6.113}$$

corresponding to case 2b for $\chi > 0$, the special version of case 1 for $\chi = 0$, and case 2a for $\chi < 0$. Isn't this fun?

> Work out the peak classifications for the SE interferogram at $t_{CB} = -\tau_v$ plotted in Fig. 6.11. For this value of the waiting time, the peaks are diagonally elongated and centered where both t_{BA} and t_{DC} are integer multiples of τ_v.

It is possible to illuminate further the form of the difference-phased SE interferogram of the simple system under consideration. For instance, the *signs* of the integer-period peaks in Fig. 6.10 can be rationalized using either the cumulant-expansion expression for the relevant overlap or a formula employing creation and annihilation operators for vibrational quanta in the g- and e-state harmonic potentials. In the latter approach, we write

$$\begin{aligned} e^{i\bar{\Lambda}(t_{DC}-t_{BA})}\langle 0_g|[-t_{CA}]_{ee}[-t_{DC}]_{gg}[t_{DB}]_{ee}[t_{BA}]_{gg}|0_g\rangle = e^{i\omega\delta^2(t_{DC}-t_{BA})} \\ \times \langle 0_g|e^{i\omega t_{BA}a_e^\dagger a_e}e^{i\omega t_{CB}a_e^\dagger a_e}e^{i\omega t_{DC}a_g^\dagger a_g}e^{-i\omega t_{DC}a_e^\dagger a_e}e^{-i\omega t_{CB}a_e^\dagger a_e}e^{-i\omega t_{BA}a_g^\dagger a_g}|0_g\rangle , \end{aligned} \tag{6.114}$$

with $a_e = a_g - \delta$ and $a_e^\dagger = a_g^\dagger - \delta$, as in the exercise on page 90. For integer-period t_{BA} and t_{DC}, along with $t_{CB} = 2.5\tau_v$, eqn (6.114) simplifies to

$$e^{i\omega\delta^2(t_{DC}-t_{BA})}\langle 0_g|1^{a_e^\dagger a_e}(-1)^{a_e^\dagger a_e}1^{a_g^\dagger a_g}1^{a_e^\dagger a_e}(-1)^{a_e^\dagger a_e}1^{a_g^\dagger a_g}|0_g\rangle = e^{i\omega\delta^2(t_{DC}-t_{BA})} ; \tag{6.115}$$

the value $e^{2\pi i\delta^2} = -0.930 + 0.368i$ explains the near sign-change seen in the interferogram when an integer-period t_{BA} is advanced by τ_v while an integer-period t_{DC} is held constant, or vice versa.

> Examine analogously the relative signs of the peaks in Fig. 6.10 where both t_{BA} and t_{DC} are odd half-integer multiples of the vibrational period τ_v.

The peaks observed in Fig. 6.10 around integer values of both t_{BA}/τ_v and t_{DC}/τ_v display interesting oscillations along t_{DC}. The argument $\omega\delta^2 t = 2\pi\delta^2 t/\tau_v$ gives the aliasing factor a period $\tau_v \delta^{-2} = 0.694$; this is too long to be the source of those oscillations, which in any case occur only with varying t_{DC}. To find their source, let's evaluate the derivative of the wave-packet overlap:

$$\frac{\partial}{\partial t_{DC}} e^{i\bar{\Lambda}(t_{DC}-t_{BA})} \langle 0_g | [-t_{CA}]_{ee} [-t_{DC}]_{gg} [t_{DB}]_{ee} [t_{BA}]_{gg} | 0_g \rangle \tag{6.116}$$
$$= \frac{i}{\hbar} e^{i\bar{\Lambda}(t_{DC}-t_{BA})} \langle 0_g | [-t_{CA}]_{ee} [-t_{DC}]_{gg} (\hbar\omega\delta^2 + H_g - H_e) [t_{DB}]_{ee} [t_{BA}]_{gg} | 0_g \rangle \,;$$

this expression relates the rate of change of the signal to the difference between the g and e potentials at the wave packets' common location at the arrival time of the D pulse—in this case the outer turning point of excited-state motion. For the interpulse delays $t_{BA} = m\tau_v$, $t_{DC} = n\tau_v$, and $t_{CB} = 2.5\tau_v$ of Fig. 6.10, the derivative (6.116) takes the value

$$i\omega e^{2\pi i\delta^2(n-m)} \langle 0_g | (-1)^{a_e^\dagger a_e} (\delta^2 + a_g^\dagger a_g - a_e^\dagger a_e)(-1)^{a_e^\dagger a_e} | 0_g \rangle = 8\pi i \frac{\delta^2}{\tau_v} e^{2\pi i \delta^2 (n-m)} \,. \tag{6.117}$$

This equation predicts an oscillation period $\tau_v/(4\delta^2) = 0.174\tau_v$, which is similar to that seen in Fig. 6.10.

> Let $t_{CB} = 2.5\tau_v$, $t_{BA} = m\tau_v$, and $t_{DC} = (n+\nu)\tau_v$. Find and re-exponentiate an expansion for the short-pulse, difference-phased SE overlap through second order in ν, verifying the prediction (6.117) for the oscillation frequency near the dual integer-period peaks in Fig. 6.10 and also estimating the FWe^{-1}M of these peaks in the t_{DC} direction.

As a step toward more complicated dynamics, we add to our 1D displaced-harmonic system a second, slower vibrational mode that mimics time-dependent solvation. Using its root-mean-squared position width $\langle q_\lambda^2 \rangle^{1/2} = \sqrt{\hbar/2\omega_\lambda}$ and its dimensionless e-state displacement $\delta_\lambda = d_\lambda \sqrt{\omega_\lambda/2\hbar}$, we can infer from the reflection principle that this mode broadens absorption lines by an amount $\hbar\omega_\lambda\delta_\lambda$. In dynamical terms, we can state that due to its momentum acquisition after an electronic transition, the λ-mode momentum reaches its root-mean-squared value, leading to a loss of overlap between the evolving excited-state wave packet and its ground-state parent after $\sim 1/2\omega_\lambda\delta_\lambda$. During this "optical dephasing time" the coordinate changes by about $\frac{1}{8\delta_\lambda}\sqrt{\frac{2\hbar}{\omega_\lambda}}$, lowering the instantaneous vertical electronic transition energy by $\hbar\omega_\lambda/4$.

We choose $\delta_\lambda = 4$ so that the slow mode's contribution to the linewidth exceeds the Stokes shifting it generates by the optical dephasing time. The broadening induced by this mode can then be regarded as inhomogeneous. For the purpose of illustration, we'd like the dephasing time roughly to match the period $\tau_v = 2\pi/\omega$ of the higher-frequency vibration; so we're led to choose a frequency $\omega_\lambda = \omega/16\pi$.

> Verify the physical arguments leading to the chosen values of δ_λ and ω_λ.

Since the high- and low-frequency modes are assumed to be uncoupled, and the initial state is taken to be a tensor product of their respective vibrational ground

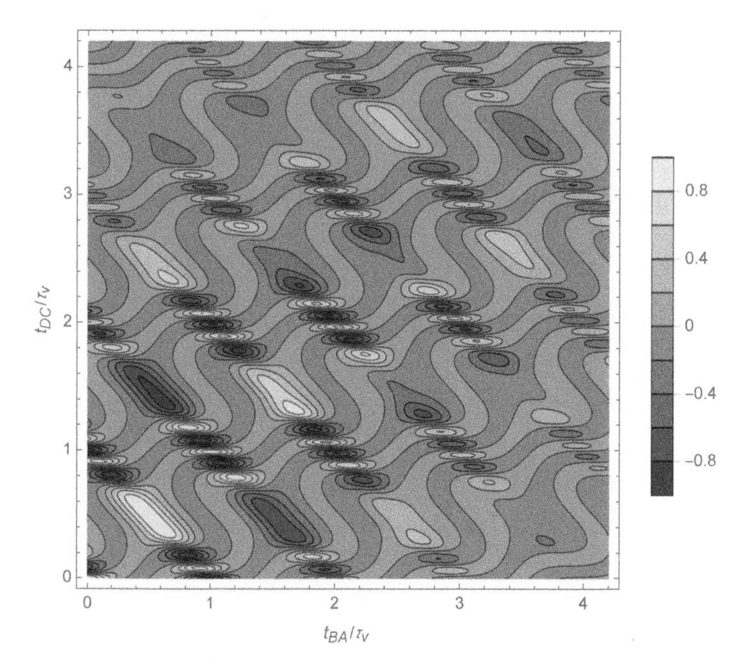

Fig. 6.13 Real part of overlap determining SE contribution to $P^{(d)}$ for two-dimensional harmonic system with displacements $\delta = 1.2$ and $\delta_\lambda = 4.0$, calculated in the short-pulse limit at $t_{CB} = 2.5\,\tau_v$.

states, the wave-packet overlaps determining the difference-phased WPI signal are given by expressions (6.101) through (6.103) with each g comprising a sum of terms for the two separate modes. Figure 6.13 shows the real part of the exponential function determining the SE overlap at $t_{CB} = 2.5\tau_v$, aliased by a factor $\exp\{i\bar{\Lambda}(t_{DC} - t_{BA})\}$ with $\bar{\Lambda}$ now equal to $\Lambda + \omega\delta^2 + \omega_\lambda\delta_\lambda^2$. In contrast to the corresponding overlap for a one-mode system shown in Fig. 6.10, this interferogram is seen to decay due to inhomogeneous broadening with increasing t_{BA} or t_{DC}; but it decays less rapidly when both intrapulse-pair delays increase in concert.

It is possible to characterize these decays by expanding in powers of the short delays t_{BA} and t_{DC} (and then re-exponentiating) the additional factor contributing to the SE overlap due to the presence of the low-frequency mode:

$$
\begin{aligned}
&e^{i\omega_\lambda\delta_\lambda^2(t_{DC}-t_{BA})}\langle 0_{\lambda g}|e^{i\omega_\lambda t_{CA}a_{\lambda e}^\dagger a_{\lambda e}}e^{i\omega_\lambda t_{DC}a_{\lambda g}^\dagger a_{\lambda g}}e^{-i\omega_\lambda t_{DB}a_{\lambda e}^\dagger a_{\lambda e}}e^{-i\omega_\lambda t_{BA}a_{\lambda g}^\dagger a_{\lambda g}}|0_{\lambda g}\rangle \\
&\quad\cong \exp\big\{2i\omega_\lambda\delta_\lambda^2(1-\cos\omega_\lambda t_{CB}) - \tfrac{1}{2}\omega_\lambda^2\delta_\lambda^2 t_{BA}^2 - \tfrac{1}{2}\omega_\lambda^2\delta_\lambda^2 t_{DC}^2(1-2i\sin\omega_\lambda t_{CB}) \\
&\qquad\quad + \omega_\lambda^2\delta_\lambda^2 t_{BA}t_{DC}e^{i\omega_\lambda t_{CB}}\big\}; \qquad (6.118)
\end{aligned}
$$

$a_{\lambda e} = a_{\lambda g} - \delta_\lambda$ and $a_{\lambda e}^\dagger = a_{\lambda g}^\dagger - \delta_\lambda$ in the first member of this approximate equality.

Derive eqn (6.118) and show that it captures the signal decays seen in Fig. 6.13.

7

Two-dimensional wave-packet interferometry for an electronic energy-transfer dimer

7.1 Energy-transfer dimer

This chapter works through the basic description of phase-coherent multidimensional electronic spectroscopy measurements on a molecular dimer comprising four *site states*, $|gg\rangle$, $|eg\rangle$, $|ge\rangle$, and $|ee\rangle$, in which neither, one, or both of the monomers are electronically excited.[1,2] Such a complex can exhibit electronic energy (or excitation) transfer (EET) between the singly excited states. The dimer Hamiltonian is $H = T + H_{el}(\hat{Q})$,[3] where T is the nuclear kinetic energy. The electronic Hamiltonian in the site basis is

$$\begin{aligned} H_{el}(Q) = {} & |gg\rangle V_{gg}(Q)\langle gg| + |eg\rangle V_{eg}(Q)\langle eg| + |ge\rangle V_{ge}(Q)\langle ge| \\ & + |ee\rangle V_{ee}(Q)\langle ee| + J(Q)(|eg\rangle\langle ge| + |ge\rangle\langle eg|) \,. \end{aligned} \tag{7.1}$$

Q stands for the full collection of intramolecular and intermolecular nuclear coordinates, including those of any surrounding medium. Alternatively, the electronic Hamiltonian can be expressed in terms of the *adiabatic electronic states*:

$$\begin{aligned} H_{el}(Q) = {} & |0\rangle E_0(Q)\langle 0| + |\bar{1}(Q)\rangle E_{\bar{1}}(Q)\langle \bar{1}(Q)| \\ & + |1(Q)\rangle E_1(Q)\langle 1(Q)| + |2\rangle E_2(Q)\langle 2| \,. \end{aligned} \tag{7.2}$$

It is easy to specify the relationship between these two representations. Let

$$\mathcal{P}_{one} = |eg\rangle\langle eg| + |ge\rangle\langle ge| \,, \tag{7.3}$$

$$\sigma_x = |ge\rangle\langle eg| + |eg\rangle\langle ge| \,, \tag{7.4}$$

$$\sigma_y = i|ge\rangle\langle eg| - i|eg\rangle\langle ge| \,, \tag{7.5}$$

and

$$\sigma_z = |eg\rangle\langle eg| - |ge\rangle\langle ge| \,. \tag{7.6}$$

Introducing the following functions of nuclear coordinates,

[1]The monomers may be the same or different. Our model neglects states such as $|e'g\rangle$ or $|e'e''\rangle$ in which one or both of the molecules occupy higher lying electronic excited states.

[2]The treatment in this chapter parallels and extends that of J. A. Cina and A. J. Kiessling, "Nuclear wave-packet dynamics in two-dimensional interferograms of excitation-transfer systems," in *Coherent Multidimensional Spectroscopy*, edited by Minhaeng Cho, Springer Series in Optical Sciences (Springer Nature, Singapore, 2019).

[3]Hats are used only when it may be specifically helpful to identify quantum mechanical operators.

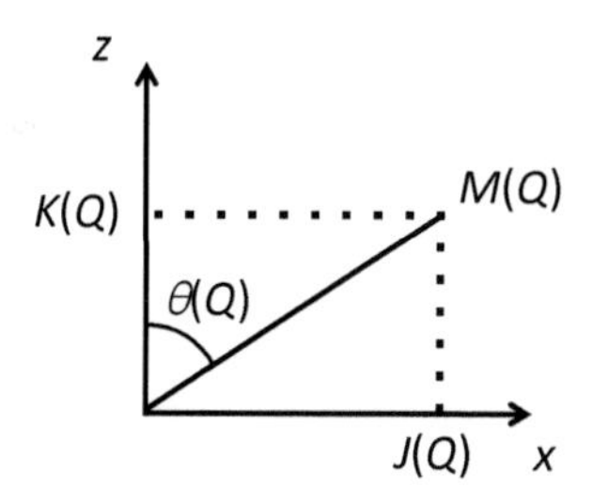

Fig. 7.1 Parameters of the electronic Hamiltonian. $J(Q)$ is the energy-transfer coupling at nuclear configuration Q. $K(Q)$ is half the local site-energy difference. $M(Q)$ is half the resultant energy difference between adiabatic singly excited electronic states.

$$K(Q) = \frac{1}{2}\left(V_{eg}(Q) - V_{ge}(Q)\right), \tag{7.7}$$

$$L(Q) = \frac{1}{2}\left(V_{eg}(Q) + V_{ge}(Q)\right), \tag{7.8}$$

$$M(Q) = \sqrt{J^2(Q) + K^2(Q)}, \tag{7.9}$$

as well as

$$\cos\theta(Q) = \frac{K(Q)}{M(Q)} \quad \text{and} \quad \sin\theta(Q) = \frac{J(Q)}{M(Q)}, \tag{7.10}$$

as illustrated in Fig. 7.1, allows us to rewrite eqn (7.1) as

$$\begin{aligned} H_{el}(Q) &= |gg\rangle\langle gg|V_{gg}(Q) + \sigma_x J(Q) + \sigma_z K(Q) + \mathcal{P}_{one}L(Q) + |ee\rangle\langle ee|V_{ee}(Q) \\ &= |gg\rangle\langle gg|V_{gg}(Q) + \mathcal{P}_{one}L(Q) \\ &\quad + e^{-i\sigma_y\theta(Q)/2}\sigma_z e^{i\sigma_y\theta(Q)/2}M(Q) + |ee\rangle\langle ee|V_{ee}(Q). \end{aligned} \tag{7.11}$$

Use the properties of the Pauli operators (7.3)–(7.6) to obtain eqn (7.11).

The adiabatic eigenenergies in eqn (7.2) are seen from eqn (7.11) to be

$$E_0(Q) = V_{gg}(Q), \tag{7.12}$$

$$E_{\bar{1}}(Q) = L(Q) + M(Q), \tag{7.13}$$

$$E_1(Q) = L(Q) - M(Q), \tag{7.14}$$

and

$$E_2(Q) = V_{ee}(Q). \tag{7.15}$$

The ground and doubly excited adiabatic eigenstates of the model dimer, $|0\rangle = |gg\rangle$ and $|2\rangle = |ee\rangle$, respectively, remain unchanged from the site basis. Singly excited adiabatic states can be defined as

$$|\bar{1}(Q)\rangle = e^{-i\sigma_y\theta(Q)/2}|eg\rangle = |eg\rangle\cos\frac{\theta(Q)}{2} + |ge\rangle\sin\frac{\theta(Q)}{2}, \tag{7.16}$$

and

$$|1(Q)\rangle = e^{-i\sigma_y\theta(Q)/2}|ge\rangle = -|eg\rangle \sin\frac{\theta(Q)}{2} + |ge\rangle\cos\frac{\theta(Q)}{2}\,. \tag{7.17}$$

The electronic Hamiltonian for this dimer has been diagonalized by a Q-dependent unitary transformation; but the remaining contribution to the full Hamiltonian, the nuclear kinetic-energy operator T, may couple the adiabatic electronic states of the single-excitation manifold in the presence of sufficiently rapid nuclear motion or approximate degeneracy between some of their vibronic levels.[4] We shall see shortly that the dipole moment operator connecting single-excitation states to the electronic ground state or the doubly excited state becomes Q-dependent in the basis of adiabatic electronic states; this feature complicates the description (using pulse propagators) of short-pulse-driven electronic transitions. Such complications do not arise in the site basis, with its nuclear coordinate-independent electronic states.

Another popular basis featuring Q-independent electronic states is the *exciton basis*. It consists of the four eigenstates of the electronic Hamiltonian evaluated at the equilibrium nuclear configuration ($Q = 0$) of the electronic ground state: $|0\rangle$, $|1\rangle \equiv |1(0)\rangle$, $|\bar{1}\rangle \equiv |\bar{1}(0)\rangle$, and $|2\rangle$.[5] The matrix elements of $H_{el}(Q)$ in this basis are readily obtained. In particular,

$$\begin{aligned}\langle\bar{1}|H_{el}(Q)|\bar{1}\rangle &= \langle\bar{1}|\bar{1}(Q)\rangle\langle\bar{1}(Q)|\bar{1}\rangle E_{\bar{1}}(Q) + \langle\bar{1}|1(Q)\rangle\langle 1(Q)|\bar{1}\rangle E_1(Q)\\ &= \langle eg|e^{-i\frac{\delta\theta(Q)}{2}\sigma_y}|eg\rangle\langle eg|e^{i\frac{\delta\theta(Q)}{2}\sigma_y}|eg\rangle E_{\bar{1}}(Q)\\ &\quad + \langle eg|e^{-i\frac{\delta\theta(Q)}{2}\sigma_y}|ge\rangle\langle ge|e^{i\frac{\delta\theta(Q)}{2}\sigma_y}|eg\rangle E_1(Q)\,,\end{aligned} \tag{7.18}$$

with $\delta\theta(Q) \equiv \theta(Q) - \theta$. Simplification leads to

$$\langle\bar{1}|H_{el}(Q)|\bar{1}\rangle = L(Q) + M(Q)\cos\delta\theta(Q)\,. \tag{7.19}$$

By similar analyses,

$$\langle 1|H_{el}(Q)|1\rangle = L(Q) - M(Q)\cos\delta\theta(Q)\,, \tag{7.20}$$

and

$$\langle 1|H_{el}(Q)|\bar{1}\rangle = \langle\bar{1}|H_{el}(Q)|1\rangle = M(Q)\sin\delta\theta(Q)\,. \tag{7.21}$$

Verify the derivation of eqns (7.19)–(7.21).

Using the definition of $\delta\theta(Q)$, we find

$$\cos\delta\theta(Q) = \frac{K(Q)K + J(Q)J}{M(Q)M}\,, \tag{7.22}$$

and

[4]See D. M. Jonas, "Vibrational and nonadiabatic coherence in 2D electronic spectroscopy, the Jahn-Teller effect, and energy transfer," Annu. Rev. Phys. Chem. **69**, 327–352 (2018).

[5]S. J. Jang, *Dynamics of Molecular Excitons* (Elsevier, Amsterdam, 2020).

$$\sin \delta\theta(Q) = \frac{J(Q)K - K(Q)J}{M(Q)M} \,. \tag{7.23}$$

Combining these various results and making the definitions $\delta J(Q) = J(Q) - J$ and $\delta K(Q) = K(Q) - K$, we arrive at

$$\begin{aligned} H_{el}(Q) &= |0\rangle\langle 0| E_0(Q) + |\bar{1}\rangle\langle\bar{1}| \Big\{ L(Q) + M + \frac{K\delta K(Q) + J\delta J(Q)}{M} \Big\} \\ &+ |1\rangle\langle 1| \Big\{ L(Q) - M - \frac{K\delta K(Q) + J\delta J(Q)}{M} \Big\} \\ &+ (|\bar{1}\rangle\langle 1| + |1\rangle\langle\bar{1}|) \Big\{ \frac{K\delta J(Q) - J\delta K(Q)}{M} \Big\} + |2\rangle\langle 2| E_2(Q) \,, \end{aligned} \tag{7.24}$$

for the electronic Hamiltonian in the exciton basis.

In order to choose between the two nuclear coordinate-independent electronic bases in a given instance, we can compare the respective ratios of the off-diagonal matrix elements of their Hamiltonians to the difference between the two diagonal matrix elements within the singly excited manifold. The site basis is favored by small values over most of the range of nuclear motion of the quantity

$$\left| \frac{J(Q)}{V_{eg}(Q) - V_{ge}(Q)} \right| = \frac{1}{2} \left| \frac{J(Q)}{K(Q)} \right| . \tag{7.25}$$

On the other hand, small values of

$$\left| \frac{K\delta J(Q) - J\delta K(Q)}{M} \frac{1}{2\big(M + \frac{K\delta K(Q) + J\delta J(Q)}{M}\big)} \right| = \frac{1}{2} \left| \frac{K\delta J(Q) - J\delta K(Q)}{M^2 + K\delta K(Q) + J\delta J(Q)} \right| \tag{7.26}$$

support a choice of the exciton basis. These criteria indicate that the site (exciton) basis is preferred under conditions of weak (strong) energy-transfer coupling. But there are special circumstances under which these two bases do not differ greatly.

> Identify parameter conditions where the quantities referred to in both eqns (7.25) and (7.26) are much less than one, and the site and exciton bases nearly coincide.

7.2 Whoopee signal

7.2.1 Interaction Hamiltonian

Next, we develop basic formulas for the fluorescence-detected WPI signal from a dimer complex. Analogous expressions can be found for the FWM signal (see Chapter 6). In order to carry out this derivation, we add to the Hamiltonian a time-dependent perturbation,

$$V(t) = \sum_{I=A,B,C,D} V_I(t) \,, \tag{7.27}$$

with

$$V_I(t) = -\hat{\mathbf{m}} \cdot \mathbf{E}_I(t) \,, \tag{7.28}$$

accounting for the dimer's interaction with a sequence of ultrashort laser pulses whose electric fields are

$$\mathbf{E}_I(t) = \mathbf{e}_I E_I f_I\left(t - t_I(\mathbf{r})\right) \cos\left[\Omega(t - t_I(\mathbf{r})) + \varphi_I\right], \tag{7.29}$$

in which $t_I(\mathbf{r}) = t_I + \mathbf{n}_I \cdot \mathbf{r}/c$. The pulse arrivals at the sample origin obey $t_A \le t_B$, $t_C \le t_D$, and $t_A + t_B \le t_C + t_D$. Here, the dipole operator is

$$\hat{\mathbf{m}} = \mathbf{m}_a\left(|eg\rangle\langle gg| + |ee\rangle\langle ge|\right) + \mathbf{m}_b\left(|ge\rangle\langle gg| + |ee\rangle\langle eg|\right) + \text{H.c.} \tag{7.30}$$

We can find this operator in the adiabatic electronic basis by calculating its matrix elements,

$$\begin{aligned}\langle\bar{1}(Q)|\hat{\mathbf{m}}|0\rangle &= \mathbf{m}_a\langle\bar{1}(Q)|eg\rangle + \mathbf{m}_b\langle\bar{1}(Q)|ge\rangle \\ &= \mathbf{m}_a \cos\frac{\theta(Q)}{2} + \mathbf{m}_b \sin\frac{\theta(Q)}{2},\end{aligned} \tag{7.31}$$

$$\langle 1(Q)|\hat{\mathbf{m}}|0\rangle = -\mathbf{m}_a \sin\frac{\theta(Q)}{2} + \mathbf{m}_b \cos\frac{\theta(Q)}{2}, \tag{7.32}$$

$$\langle 2|\hat{\mathbf{m}}|\bar{1}(Q)\rangle = \mathbf{m}_a \sin\frac{\theta(Q)}{2} + \mathbf{m}_b \cos\frac{\theta(Q)}{2}, \tag{7.33}$$

and

$$\langle 2|\hat{\mathbf{m}}|1(Q)\rangle = \mathbf{m}_a \cos\frac{\theta(Q)}{2} - \mathbf{m}_b \sin\frac{\theta(Q)}{2}. \tag{7.34}$$

From these matrix elements, we find

$$\begin{aligned}\hat{\mathbf{m}} = |\bar{1}(Q)\rangle\langle 0|&\Big\{\mathbf{m}_a \cos\frac{\theta(Q)}{2} + \mathbf{m}_b \sin\frac{\theta(Q)}{2}\Big\} \\ + |1(Q)\rangle\langle 0|&\Big\{-\mathbf{m}_a \sin\frac{\theta(Q)}{2} + \mathbf{m}_b \cos\frac{\theta(Q)}{2}\Big\} \\ + |2\rangle\langle\bar{1}(Q)|&\Big\{\mathbf{m}_a \sin\frac{\theta(Q)}{2} + \mathbf{m}_b \cos\frac{\theta(Q)}{2}\Big\} \\ + |2\rangle\langle 1(Q)|&\Big\{\mathbf{m}_a \cos\frac{\theta(Q)}{2} - \mathbf{m}_b \sin\frac{\theta(Q)}{2}\Big\} + \text{H.c.}\end{aligned} \tag{7.35}$$

Setting $Q = 0$ in eqn (7.35) yields the dipole operator in the exciton basis:

$$\begin{aligned}\hat{\mathbf{m}} = |\bar{1}\rangle\langle 0|\Big\{\mathbf{m}_a \cos\frac{\theta}{2} + \mathbf{m}_b \sin\frac{\theta}{2}\Big\} + |1\rangle\langle 0|\Big\{-\mathbf{m}_a \sin\frac{\theta}{2} + \mathbf{m}_b \cos\frac{\theta}{2}\Big\} \\ + |2\rangle\langle\bar{1}|\Big\{\mathbf{m}_a \sin\frac{\theta}{2} + \mathbf{m}_b \cos\frac{\theta}{2}\Big\} + |2\rangle\langle 1|\Big\{\mathbf{m}_a \cos\frac{\theta}{2} - \mathbf{m}_b \sin\frac{\theta}{2}\Big\} + \text{H.c.}.\end{aligned} \tag{7.36}$$

7.2.2 2D Signal

The two-dimensional whoopee signal depends on the quadrilinear contributions (proportional to $E_A E_B E_C E_D$) to each of the several electronic excited states. For a sample having dimer density ρ throughout an illuminated volume V, it can be written

$$\begin{aligned} S &= \rho \int_V d^3r \{ Q_{ee} P_{ee}(\mathbf{r}) + Q_{eg} P_{eg}(\mathbf{r}) + Q_{ge} P_{ge}(\mathbf{r}) \} \\ &= \rho \int_V d^3r \{ Q_2 P_2(\mathbf{r}) + Q_{\bar{1}} P_{\bar{1}}(\mathbf{r}) + Q_1 P_1(\mathbf{r}) \}, \end{aligned} \tag{7.37}$$

where $P_\xi(\mathbf{r})$ is the quadrilinear contribution to the ξ-state population of a dimer at position $\mathbf{r}$ and Q_ξ is the fluorescence quantum yield in that state.[6] The $\mathbf{r}$-dependence of the quadrilinear populations would become important if we wished to describe a set-up in which the four incident laser beams are noncollinear; in this case, wave-vector-matching conditions among the $\mathbf{n}_I$ can play a role analogous to or in concert with optical phase cycling in helping to isolate a particular quadrilinear contribution to the signal (i.e. one having a sum or difference optical phase combination, $\varphi_{BA}+\varphi_{DC}$ or $\varphi_{BA} - \varphi_{DC}$, respectively). In order to simplify subsequent formulas, we shall omit the $\mathbf{r}$-dependence from here on out. The resulting expressions will apply as they are in the case of four collinear beams or in the case of experiments on a single dimer, with the spatial integrals of eqn (7.37) also being omitted in the latter instance.

Omitting from the 48 quadrilinear contributions to the population of a given singly-excited state the sixteen overlaps carrying an uncontrolled optical phase, we can break the remaining P_ξ into separately measurable sum- and difference-phased components. Thus $P_\xi = P_\xi^{(s)} + P_\xi^{(d)}$, where

$$\begin{aligned} P_\xi^{(s)} = 2\mathrm{Re}\{ & \langle \uparrow_A \downarrow_B \uparrow_C |\xi\rangle\langle\xi| \uparrow_D\rangle + \langle \uparrow_A \downarrow_D \uparrow_C |\xi\rangle\langle\xi| \uparrow_B\rangle + \langle \uparrow_C |\xi\rangle\langle\xi| \uparrow_B \downarrow_A \uparrow_D\rangle \\ &+ \langle \uparrow_A |\xi\rangle\langle\xi| \uparrow_B \downarrow_C \uparrow_D\rangle + \langle \uparrow_C \downarrow_B \uparrow_A |\xi\rangle\langle\xi| \uparrow_D\rangle + \langle \uparrow_C \downarrow_D \uparrow_A |\xi\rangle\langle\xi| \uparrow_B\rangle \\ &+ \langle \uparrow_C |\xi\rangle\langle\xi| \uparrow_D \downarrow_A \uparrow_B\rangle + \langle \uparrow_A |\xi\rangle\langle\xi| \uparrow_D \downarrow_C \uparrow_B\rangle \\ &+ \langle \uparrow_A \uparrow_C \downarrow_B |\xi\rangle\langle\xi| \uparrow_D\rangle + \langle \uparrow_A \uparrow_C \downarrow_D |\xi\rangle\langle\xi| \uparrow_B\rangle + \langle \uparrow_C |\xi\rangle\langle\xi| \uparrow_B \uparrow_D \downarrow_A\rangle \\ &+ \langle \uparrow_A |\xi\rangle\langle\xi| \uparrow_B \uparrow_D \downarrow_C\rangle + \langle \uparrow_C \uparrow_A \downarrow_B |\xi\rangle\langle\xi| \uparrow_D\rangle + \langle \uparrow_C \uparrow_A \downarrow_D |\xi\rangle\langle\xi| \uparrow_B\rangle \\ &+ \langle \uparrow_C |\xi\rangle\langle\xi| \uparrow_D \uparrow_B \downarrow_A\rangle + \langle \uparrow_A |\xi\rangle\langle\xi| \uparrow_D \uparrow_B \downarrow_C\rangle \}, \end{aligned} \tag{7.38}$$

and (by swapping Cs and Ds throughout)

$$\begin{aligned} P_\xi^{(d)} = 2\mathrm{Re}\{ & \langle \uparrow_A \downarrow_B \uparrow_D |\xi\rangle\langle\xi| \uparrow_C\rangle + \langle \uparrow_A \downarrow_C \uparrow_D |\xi\rangle\langle\xi| \uparrow_B\rangle + \langle \uparrow_D |\xi\rangle\langle\xi| \uparrow_B \downarrow_A \uparrow_C\rangle \\ &+ \langle \uparrow_A |\xi\rangle\langle\xi| \uparrow_B \downarrow_D \uparrow_C\rangle + \langle \uparrow_D \downarrow_B \uparrow_A |\xi\rangle\langle\xi| \uparrow_C\rangle + \langle \uparrow_D \downarrow_C \uparrow_A |\xi\rangle\langle\xi| \uparrow_B\rangle \\ &+ \langle \uparrow_D |\xi\rangle\langle\xi| \uparrow_C \downarrow_A \uparrow_B\rangle + \langle \uparrow_A |\xi\rangle\langle\xi| \uparrow_C \downarrow_D \uparrow_B\rangle \\ &+ \langle \uparrow_A \uparrow_D \downarrow_B |\xi\rangle\langle\xi| \uparrow_C\rangle + \langle \uparrow_A \uparrow_D \downarrow_C |\xi\rangle\langle\xi| \uparrow_B\rangle + \langle \uparrow_D |\xi\rangle\langle\xi| \uparrow_B \uparrow_C \downarrow_A\rangle \\ &+ \langle \uparrow_A |\xi\rangle\langle\xi| \uparrow_B \uparrow_C \downarrow_D\rangle + \langle \uparrow_D \uparrow_A \downarrow_B |\xi\rangle\langle\xi| \uparrow_C\rangle + \langle \uparrow_D \uparrow_A \downarrow_C |\xi\rangle\langle\xi| \uparrow_B\rangle \\ &+ \langle \uparrow_D |\xi\rangle\langle\xi| \uparrow_C \uparrow_B \downarrow_A\rangle + \langle \uparrow_A |\xi\rangle\langle\xi| \uparrow_C \uparrow_B \downarrow_D\rangle \}. \end{aligned} \tag{7.39}$$

[6] We do not include an average over any possible orientational distribution of the dimer at a given location, regarding this as being implicitly included in the spatial distribution of population.

Explicit expressions for the contributing multi-pulse amplitudes are developed subsequently. We have written the various quadrilinear populations so the displayed overlaps have phase signature $e^{-i\varphi_{BA}-i\varphi_{DC}}$ and $e^{-i\varphi_{BA}+i\varphi_{DC}}$ in $P_{\xi}^{(s)}$ and $P_{\xi}^{(d)}$, respectively. The first eight overlaps in each of the quadrilinear ξ-state populations are in direct correspondence with those of eqns (6.9) and (6.10) in Chapter 6. They do not access the doubly excited electronic state at any stage in the amplitude-transfer process described by the three-pulse bra or ket. Amplitude in the doubly excited state is generated by the action of the second pulse of the three-pulse bra or ket in the last eight overlaps in $P_{\xi}^{(s)}$ and $P_{\xi}^{(d)}$, which have no counterparts in the two-state monomer model.

If we make the reasonable assumption that the quantum yields for fluorescence (or other action-spectroscopy) signal following excitation of the *eg*- and *ge*-states are equal, $Q_{eg} = Q_{ge} = Q_{one}$, then the total quadrilinear population of both singly-excited site states $P_{one}^{(s/d)} = P_{eg}^{(s/d)} + P_{ge}^{(s/d)}$ is all that matters. Expressions for $P_{one}^{(s)}$ and $P_{one}^{(d)}$ could be obtained from eqns (7.38) and (7.39), respectively by replacing $|\xi\rangle\langle\xi|$ with $\mathcal{P}_{one} = |eg\rangle\langle eg| + |ge\rangle\langle ge|$ (see eqn (7.3)). But all the one- and three-pulse bras and kets appearing in eqns (7.38) and (7.39) generate amplitude only in the singly electronically excited manifold. So the operator $|\xi\rangle\langle\xi|$ can simply be removed to give expressions for $P_{one}^{(s/d)}$. The same sort of argument applies in the exciton basis.

The region of interpulse-delay space $\{t_{BA}, t_{DC}, t_{CB}\}$ containing each of the first eight overlaps in $P_{one}^{(s)}$ and $P_{one}^{(d)}$ is the same as that identified in Chapter 6 for the corresponding overlap of $P^{(s)}$ or $P^{(d)}$, respectively. Each of the last eight overlaps in eqns (7.38) and (7.39), those involving a three-pulse bra or ket whose amplitude visits the *ee*-state between the second- and third-acting pulse, can also be confined to unique regions of $\{t_{BA}, t_{DC}, t_{CB}\}$. The delay-space volumes for the second set of eight quadrilinear overlaps contributing to $P_{one}^{(d)}$ are shown in Appendix B.

It is worth recalling that outside its prescribed interpulse-delay region, a particular overlap vanishes simply because the delay combinations are inconsistent with the required order of pulse action in the participating three-pulse bra or ket. But criteria specific to the dimer's Hamiltonian, including internal molecular and host-medium nuclear degrees of freedom, will sometimes diminish the overlaps for certain delay combinations inside these regions. In multimode systems for instance, electronic dephasing driven by electronic-nuclear coupling will often severely limit the maximal intrapulse-pair delays t_{BA} and t_{DC} over which the electronic coherence on which a nonvanishing 2D-WPI signal depends can be effectively maintained. Because the *B*- and *C*-pulses are not phase-locked, signal contributions tend to be less susceptible to dynamical truncation along t_{CB}.

The last point can be illustrated by revisiting the basic 2D-WPI expressions of eqns (6.9) and (6.10) in the preceding chapter. From those formulas, we see that both bra and ket in the SE overlaps evolve in the *e*-state during t_{CB}, while in the GSB overlaps, both bra and ket evolve in the *g*-state during this interval; $P^{(s)}$ and $P^{(d)}$ are therefore expected to have decreased sensitivity to electronic decoherence during the "waiting time" t_{CB}.

Draw the delay-space volumes within $\{t_{BA}, t_{DC}, t_{CB}\}$ to which the *ee*-visiting overlaps contributing to $P_{one}^{(s)}$ are confined.

As indicated in eqn (7.37), the quadrilinear portion of the population of the doubly excited electronic state contributes to the 2D-WPI signal with a certain quantum yield $Q_{ee} = Q_2 \equiv Q_{two}$. Eliminating as unmeasured the four overlaps appearing in the doubly excited population whose optical phases are uncontrolled allows us to write P_{two} as a sum $P_{two}^{(s)} + P_{two}^{(d)}$ of two experimentally isolable portions,

$$P_{two}^{(s)} = 2\mathrm{Re}\big\{\langle\uparrow_A\uparrow_C|\uparrow_B\uparrow_D\rangle + \langle\uparrow_A\uparrow_C|\uparrow_D\uparrow_B\rangle + \langle\uparrow_C\uparrow_A|\uparrow_B\uparrow_D\rangle + \langle\uparrow_C\uparrow_A|\uparrow_D\uparrow_B\rangle\big\}, \quad (7.40)$$

and (by interchanging Cs and Ds)

$$P_{two}^{(d)} = 2\mathrm{Re}\big\{\langle\uparrow_A\uparrow_D|\uparrow_B\uparrow_C\rangle + \langle\uparrow_A\uparrow_D|\uparrow_C\uparrow_B\rangle + \langle\uparrow_D\uparrow_A|\uparrow_B\uparrow_C\rangle + \langle\uparrow_D\uparrow_A|\uparrow_C\uparrow_B\rangle\big\}. \quad (7.41)$$

The regions of interpulse delay where each of the quadrilinear overlaps in $P_{two}^{(d)}$ is not prevented from vanishing by the order of pulse action is shown in Appendix C.

Determine the delay-space volume within which each of the overlaps in eqn (7.40) can be nonzero.

7.2.3 One-, two-, and three-pulse kets

Using time-dependent perturbation theory, we now seek explicit expressions for the various multi-pulse bras and kets whose overlaps determine the 2D-WPI signal from an EET complex. The early steps in this derivation recapitulate those taken in previous chapters, but we start over from scratch nonetheless. Under the dimer Hamiltonian $H = T + H_{el}(\hat{Q})$ (see eqn (7.1)) and the interaction potential $V(t)$, given in eqn (7.27) with the pulse arrival times reckoned at $\mathbf{r} = 0$, the quantum mechanical state obeys

$$i\hbar\frac{\partial}{\partial t}|\Psi(t)\rangle = \big(H + V(t)\big)|\Psi(t)\rangle\,. \quad (7.42)$$

The initial condition is taken to be $|\Psi(t \ll t_A)\rangle = [t - t_A]|gg\rangle|\psi_0\rangle$ (using $[t] \equiv \exp\{-iHt/\hbar\}$), where $|\psi_0\rangle$ is some eigenket of the ground-state nuclear Hamiltonian $T + \langle gg|H_{el}(\hat{Q})|gg\rangle$. In the interaction picture, eqn (7.42) becomes

$$i\hbar\frac{\partial}{\partial t}|\tilde{\Psi}(t)\rangle = \tilde{V}(t))|\tilde{\Psi}(t)\rangle\,, \quad (7.43)$$

with $\tilde{V}(t) = [-t + t_A]V(t)[t - t_A]$ and $|\tilde{\Psi}(t \ll t_A)\rangle = |gg\rangle|\psi_0\rangle$.

Since the quadrilinear signal contributions all take the form of an overlap between a two-pulse bra and a two-pulse ket, or between a one-pulse bra and a three-pulse ket, it is enough to solve eqn (7.43) through third order in the external fields:

$$|\tilde{\Psi}(t)\rangle \cong \left\{1 + \frac{1}{i\hbar}\int_{-\infty}^{t} d\tau\,\tilde{V}(\tau) + \left(\frac{1}{i\hbar}\right)^2\int_{-\infty}^{t} d\tau\int_{-\infty}^{\tau} d\bar{\tau}\,\tilde{V}(\tau)\tilde{V}(\bar{\tau})\right.$$
$$\left. + \left(\frac{1}{i\hbar}\right)^3\int_{-\infty}^{t} d\tau\int_{-\infty}^{\tau} d\bar{\tau}\int_{-\infty}^{\bar{\tau}} d\bar{\bar{\tau}}\,\tilde{V}(\tau)\tilde{V}(\bar{\tau})\tilde{V}(\bar{\bar{\tau}})\right\}|gg\rangle|\psi_0\rangle\,. \quad (7.44)$$

Just as in Chapter 6, this perturbative solution can be rewritten in terms of pulse propagators that encapsulate the effect of each finite-duration laser pulse in the instantaneous action of a single quantum mechanical operator. Upon reversion to the Schrödinger picture, this reframing results in an equivalent solution,

$$\begin{aligned}|\Psi(t)\rangle = \{[t-t_A] + i\sum_{I=A,B,C,D}[t-t_I]P_I(t-t_I;\tau)[t_{IA}] \\ + i^2\sum_{IJ}[t-t_J]P_J(t-t_J;\tau)[t_{JI}]P_I(\tau+t_{JI};\bar{\tau})[t_{IA}] \\ + i^3\sum_{IJK}[t-t_K]P_K(t-t_K;\tau)[t_{KJ}]P_J(\tau+t_{KJ};\bar{\tau})[t_{JI}] \\ \times P_I(\bar{\tau}+t_{JI};\bar{\bar{\tau}})[t_{IA}]\}|gg\rangle|\psi_0\rangle\,,\end{aligned} \tag{7.45}$$

in which the Ith pulse propagator is

$$P_I(t;\tau) = \frac{E_I}{\hbar}\int_{-\infty}^{t} d\tau\, f_I(\tau)\cos(\Omega\tau+\varphi_I)\,[-\tau]\,\mathbf{e}_I\cdot\hat{\mathbf{m}}\,[\tau]\,. \tag{7.46}$$

As before, the first argument of the pulse propagator is the upper integration limit and the second designates the variable of integration.

By extracting terms from eqn (7.45) we can develop formulas for the wave-packet overlaps contributing to the 2D-WPI signal using any electronic basis. Portions of the two- and three-pulse sums in this expression of quadratic or cubic order in an individual field-strength are irrelevant. Here, we bypass the adiabatic electronic basis in favor of the two fixed bases, letting $|\xi\rangle$ denote either a site or an exciton state. Electronic matrix elements of the pulse propagator (eqn (7.46)) can be conveniently expressed in terms of reduced pulse propagators,

$$p_I^{(\xi\bar{\xi})}(t;\tau) = \int_{-\infty}^{t}\frac{d\tau}{\sigma}\, f_I(\tau)e^{\mp i\Omega\tau}[-\tau]_{\xi\xi}[\tau]_{\bar{\xi}\bar{\xi}}\,, \tag{7.47}$$

by making a rotating-wave approximation and invoking one of the forms (7.30) or (7.36) for the dipole operator. We have written $\langle\xi|[\tau]|\bar{\xi}\rangle = [\tau]_{\xi\bar{\xi}}$ and made a significant simplification by assuming that electronic transitions between different singly excited states *in the appropriate basis* can be neglected on the timescale of the pulse duration. The upper (lower) sign applies in eqn (7.47) for the case of an absorptive (emissive) transition $\xi \leftarrow \bar{\xi}$ ($\bar{\xi} \rightarrow \xi$). With the same approximations and conventions, the nonvanishing elements of the overall pulse propagator (eqn (7.46)) can be written as

$$\langle\xi|P_I(t;\tau)|\bar{\xi}\rangle = F_I^{(\xi\bar{\xi})}e^{\mp i\varphi_I}p_I^{(\xi\bar{\xi})}(t;\tau)\,, \tag{7.48}$$

where $F_I^{(\xi\bar{\xi})} \equiv F_I\langle\xi|\hat{\mathbf{m}}\cdot\mathbf{e}_I|\bar{\xi}\rangle$ with $F_I = E_I\sigma/2\hbar$.[7]

[7]Note that any polarization dependence of a 2D-WPI signal enters through the $F_I^{(\xi\bar{\xi})}$.

7.3 Illustrative calculations

7.3.1 Overlaps

The stage is set for a wide variety of calculations of WPI signals from EET dimers and their interpretation in terms of the underlying nuclear wave-packet and energy-transfer dynamics. For the purpose of illustration, we drastically pare the vast range of possible molecular parameters and experimental choices by investigating the difference-phased fluorescence-detected signal from a space-fixed dimer with perpendicularly oriented monomer transition dipoles, $\hat{\mathbf{m}}_a = m_a\hat{\mathbf{x}}$ and $\hat{\mathbf{m}}_b = m_b\hat{\mathbf{y}}$.[8,9,10] In the special case when the A, B, C, and D pulses arrive in their nominal order and are short enough that temporal pulse overlap can be neglected.[11] Consulting the relevant delay regions shown in Chapter 6 and Appendices B and C then confirms inspection of eqns (7.39) and (7.41) in showing that under these circumstances, the singly and doubly excited populations simplify to

$$P_{one}^{(d)} = 2\text{Re}\{\langle\uparrow_A\downarrow_B\uparrow_D \mid \uparrow_C\rangle + \langle\uparrow_A\downarrow_C\uparrow_D \mid \uparrow_B\rangle + \langle\uparrow_A \mid \uparrow_B\uparrow_C\downarrow_D\rangle\}\,, \tag{7.49}$$

and

$$P_{two}^{(d)} = 2\text{Re}\{\langle\uparrow_A\uparrow_D \mid \uparrow_B\uparrow_C\rangle\}\,, \tag{7.50}$$

respectively.

In a first example, we take the electronic excitation-transfer coupling to be sufficiently weak that the site states are the preferred electronic basis. Provided the interpulse delays are not too long, the condition $|Jt_{IK}|/\hbar \ll 1$ then ensures that the WPI signal will be at most of first order in J. It is easy to show that if all four laser pulses have the same (x or y) polarization, then first-order EET cannot contribute to the signal. For we can arrange the site-state labels as in Fig. 7.2, so the directions between them are those of the connecting transition moments. Four x-polarized pulses

[8]The net signal from an isotropic sample of dimers with a certain internal geometry would be a weighted sum of signals from a handful of representative space-fixed orientations.

[9]While ultrafast nonlinear optical measurements on spatially oriented samples, such as macromolecular single crystals, are not yet common, they have on occasion been successfully carried out; see, for example, L. Huang, N. Ponomarenko, G. P. Wiederrecht, and D. M. Tiede, "Cofactorspecific photochemical function resolved by ultrafast spectroscopy in photosynthetic reaction center crystals," Proc. Natl. Acad. Sci. U.S.A. **109**, 4851–4856 (2012). Experiments allowing polarization-selection-based targeting of specific chromophores within a multi-chromophore complex are likely to be aided by the advance of fluorescence-detected time-resolved experiments on single molecules.

[10]For illustrative examples of ultrafast polarization-spectroscopy measurements on *isotropic* samples, see: Y. Song, A. Schubert, E. Maret, R. K. Burdick, B. D. Dunietz, E. Geva, and J. P. Ogilvie, "Vibronic structure of photosynthetic pigments probed by polarized two-dimensional electronic spectroscopy and *ab initio* calculations," Chem. Sci. **10**, 8143–8153 (2019); N. S. Ginsberg, J. A. Davis, M. Ballottari, Y.-C. Cheng, R. Bassi, and G. R. Fleming, "Solving structure in the CP29 light harvesting complex with polarization-phased 2D electronic spectroscopy," Proc. Natl. Acad. Sci. U. S. A. **108**, 3848–3853 (2011); G. S. Schlau-Cohen, A. Ishizaki, T. R. Calhoun, N. S. Ginsberg, M. Ballottari, R. Bassi, and G. R. Fleming,"Elucidation of the timescales and origins of quantum electronic coherence in LHCII," Nat. Chem. **4**, 389–395 (2012); and R. Gera, S. L. Meloni, and J. M. Anna, "Unraveling confined dynamics of guests trapped in self-assembled Pd_6L_4 nanocages by ultrafast mid-IR polarization-dependent spectroscopy," J. Phys. Chem. Lett. **10**, 413–418 (2019).

[11]Specifically, these assumptions mean that t_{CB} is positive and greater than the pulse duration. In addition, since t_{BA} and t_{DC} are defined to be positive, they also mean that pulse-overlap effects will be ignored when either is very short.

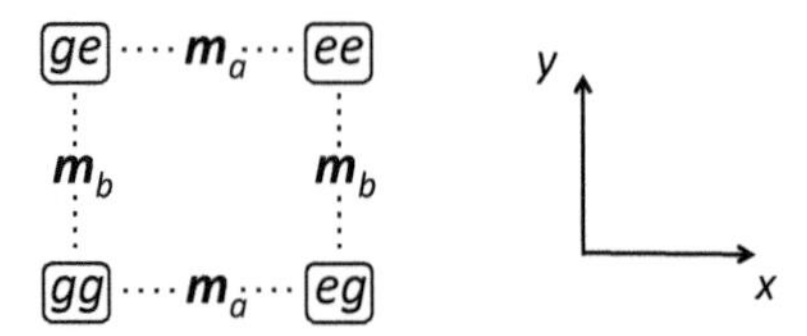

Fig. 7.2 Site-state labels arranged so that each pair is separated in the direction of the monomer transition moment that connects them.

combined with a single EET-driven amplitude transfer can then be seen to generate one- and three-pulse wave packets in *different* electronic states, as illustrated in Fig. 7.3; their overlaps vanish and hence make no contribution to P_{one}. Four x-pulses are similarly unable to contribute to P_{two} linearly in J, as Fig. 7.4 reveals.

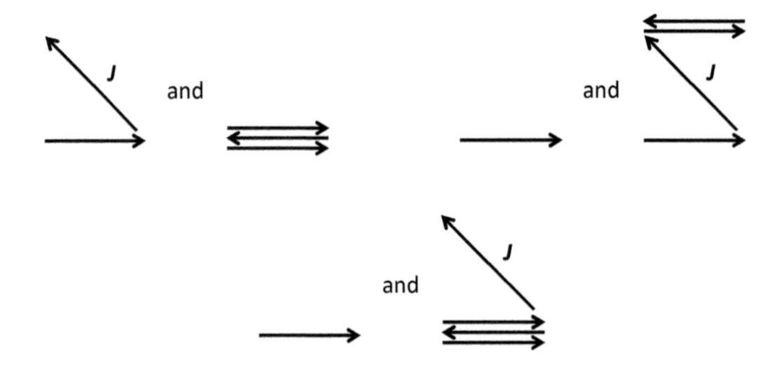

Fig. 7.3 Three combinations of one- and three-pulse states formed by all x-polarized pulses, with one state or the other being linear in J. Each of the three pairs forms a vanishing wave-packet overlap. Reprinted by permission from Springer Nature: J. A. Cina and A. J. Kiessling in *Coherent Multidimensional Spectroscopy*, edited by M. Cho (Springer Nature, Singapore, 2019).

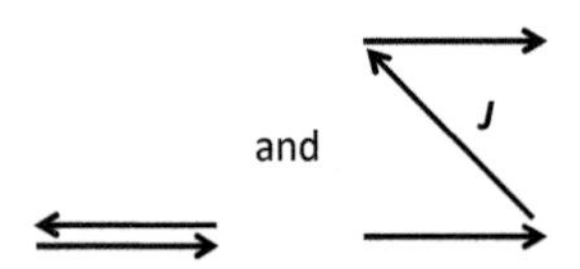

Fig. 7.4 Illustration that a pair of two-pulse wave packets, one of them linear in J, formed by four x-polarized pulses produce a vanishing overlap. Reprinted by permission from Springer Nature: J. A. Cina and A. J. Kiessling in *Coherent Multidimensional Spectroscopy*, edited by M. Cho (Springer Nature, Singapore, 2019).

As an example of a polarization combination that produces a nonzero difference-phased interference signal of first order in J from our oriented model dimer (and no signal in the absence of energy transfer), we consider the case $\mathbf{e}_A = \hat{\mathbf{y}}$, $\mathbf{e}_B = \mathbf{e}_C =$

$\mathbf{e}_D = \hat{\mathbf{x}}$.[12,13] The sequences of pulse- and EET-driven electronic transitions making up the quadrilinear overlaps of $P_{one}^{(d)}$ are sketched in Fig. 7.5, while those responsible for $P_{two}^{(d)}$ are illustrated in Fig. 7.6.[14]

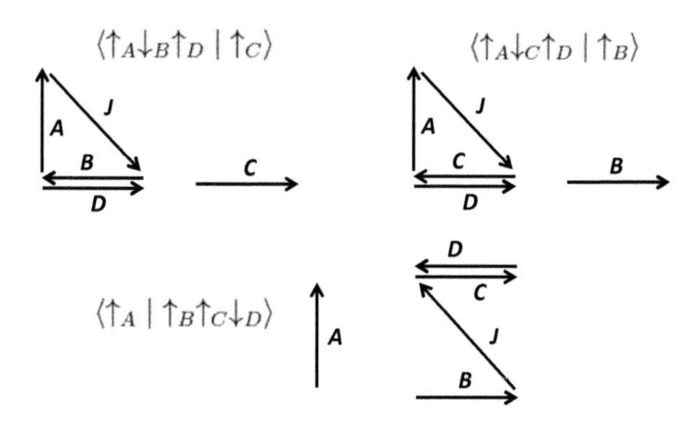

Fig. 7.5 Sequence of electronic transitions under $A_y B_x C_x D_x$ polarization in each pair of states whose overlap contributes to $P_{one}^{(d)}$ at first order in J.

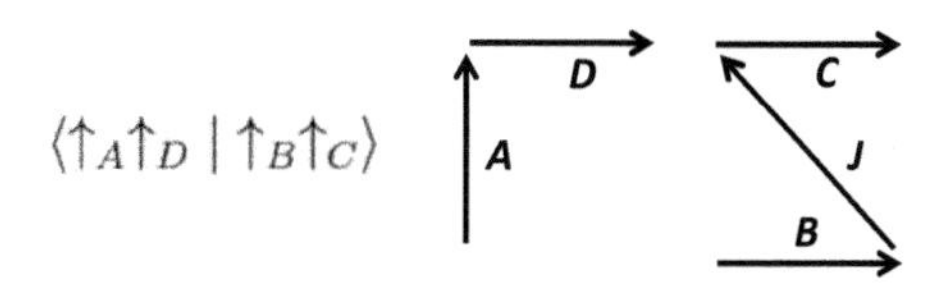

Fig. 7.6 Electronic transitions under $A_y B_x C_x D_x$ polarization in the two states whose overlap determines $P_{two}^{(d)}$ at first order in J.

We consider a simple realization of the nuclear structure in our dimer Hamiltonian (eqn (7.1)), with each monomer having a single internal vibrational mode, so the site-state potential functions are

$$V_{gg}(q_a, q_b) = \frac{m\omega^2}{2}(q_a^2 + q_b^2), \tag{7.51}$$

$$V_{eg}(q_a, q_b) = \epsilon_{eg} + \frac{m\omega^2}{2}\left((q_a - d)^2 + q_b^2\right), \tag{7.52}$$

[12] An alternative odd-polarization combination, in which $\mathbf{e}_A = \mathbf{e}_B = \mathbf{e}_C = \hat{\mathbf{x}}$ and $\mathbf{e}_D = \hat{\mathbf{y}}$, has been considered for a similar class of EET systems and shown to provide direct experimental access to the time evolution of either intersite or interexciton electronic coherence; see A. J. Kiessling and J. A. Cina, "Monitoring the evolution of intersite and interexciton coherence in electronic excitation transfer via wave-packet interferometry," J. Chem. Phys. **152**, 244311/1–19 (2020).

[13] See also S. Chatterjee and N. Makri, "Recovery of purity in dissipative tunneling dynamics," J. Phys. Chem. Lett. **11**, 8592–8596 (2020).

[14] 2D interferograms from $\langle \uparrow_A \mid \uparrow_B \uparrow_C \downarrow_D \rangle$ alone in a model dimer akin to that considered here, under the same polarization conditions but with arbitrarily abrupt laser pulses, were the subject of J. A. Cina, D. Kilin, and T. S. Humble, "Wave packet interferometry for short-time electronic energy transfer: Multidimensional optical spectroscopy in the time domain," J. Chem. Phys. **118**, 46–61 (2003).

$$V_{ge}(q_a, q_b) = \epsilon_{ge} + \frac{m\omega^2}{2}\left(q_a^2 + (q_b - d)^2\right), \tag{7.53}$$

and

$$V_{ee}(q_a, q_b) = \epsilon_{ee} + \frac{m\omega^2}{2}\left((q_a - d)^2 + (q_b - d)^2\right), \tag{7.54}$$

while the kinetic energy operator is

$$T = \frac{p_a^2}{2m} + \frac{p_b^2}{2m}. \tag{7.55}$$

We focus on a case of "downhill" energy transfer with $\epsilon_{eg} - \epsilon_{ge} = m\omega^2 d^2$, so that the intersection line between the singly excited site-state potentials passes through the minimum point $(q_a, q_b) = (d, 0)$ of the higher-energy, "donor-state" potential (see Fig. 7.7). We pick a moderate Franck–Condon displacement $d = q_{rms}$, where $q_{rms} =$

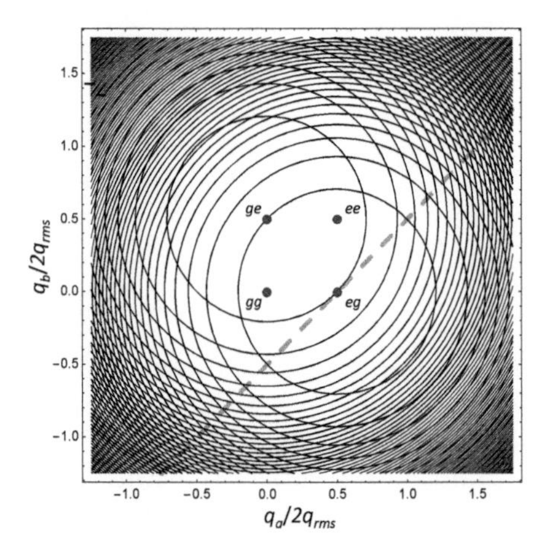

Fig. 7.7 Plots of V_{eg} and V_{ge} with their minimum locations labeled. Minima of V_{gg} and V_{ee} are also shown. Dashed diagonal is the line of intersection between V_{eg} and V_{ge}.

$\sqrt{\hbar/2m\omega}$.[15] The condition of eqn (7.25) for weak energy-transfer coupling underlying our choice of the site basis here implies $|J| \ll m\omega^2 d^2 = \hbar\omega/2$. We treat J as a constant, dismissing any nuclear-coordinate dependence.

The weak-coupling condition just invoked ensures $|Jt_{IK}/\hbar| \ll \omega t_{IK}/2$ in the present instance. So the effect of energy transfer is small and can be treated perturbatively through first order in J, as we wish to do, over interpulse delays up to several vibrational periods in length. Other than assuming that weak coupling is operative and stating J's *sign*, we need not specify its magnitude any more precisely. In the WPI signal calculations presented below the sign of J is taken positive; all the interferograms would change sign if it were instead made negative.

[15]The value $\epsilon_{eg} - \epsilon_{ge} = \hbar\omega/2$ leads to equal interleaving between the zeroth-order vibronic levels in the eg and ge states.

Next, from eqn (7.45) we extract formulas for each of the wave packets appearing in eqns (7.49) and (7.50). We find, for instance,

$$|\uparrow_C\rangle = ie^{-i\varphi_C} \sum_{\xi,\xi_1=eg,ge} |\xi\rangle F_C^{(\xi_1\, gg)}[t-t_C]_{\xi\,\xi_1}\, p_C^{(\xi_1 gg)}[t_{CA}]_{gg\,gg}|\psi_0\rangle\,, \tag{7.56}$$

and

$$|\uparrow_A\downarrow_B\uparrow_D\rangle = -ie^{i\varphi_{BA}-i\varphi_D} \sum_{\bar\xi,\xi_2,\xi_3,\xi_4=eg,ge} |\bar\xi\rangle F_D^{(\xi_2\, gg)} F_B^{(gg\,\xi_3)} F_A^{(\xi_4\, gg)}$$
$$\times [t-t_D]_{\bar\xi\,\xi_2}\, p_D^{(\xi_2\, gg)}[t_{DB}]_{gg\,gg}\, p_B^{(gg\,\xi_3)}[t_{BA}]_{\xi_3\,\xi_4}\, p_A^{(\xi_4\, gg)}|\psi_0\rangle\,. \tag{7.57}$$

Because temporal pulse overlap is being neglected, we have replaced the first argument of each reduced pulse propagator by infinity and written $p_I^{(\bar\xi\xi)}(\infty;\tau) = p_I^{(\bar\xi\xi)}$. The "observation time" t then disappears when we take the inner product between the wave packets of eqns (7.56) and (7.57), which includes a sum over the state index ξ—unitary evolution of both wave packets, including subsequent energy transfer, does not affect their contribution to the singly excited population—and we obtain

$$\langle\uparrow_A\downarrow_B\uparrow_D\mid\uparrow_C\rangle = -e^{-i\varphi_{BA}+i\varphi_{DC}} \sum_{\xi_1\,\xi_2\,\xi_3\,\xi_4} F_A^{(gg\,\xi_4)} F_B^{(\xi_3\, gg)} F_D^{(gg\,\xi_2)} F_C^{(\xi_1\, gg)} \tag{7.58}$$
$$\times \langle\psi_0|p_A^{(gg\,\xi_4)}[-t_{BA}]_{\xi_4\,\xi_3}\, p_B^{(\xi_3\, gg)}[-t_{DB}]_{gg\,gg}\, p_D^{(gg\,\xi_2)}[t_{DC}]_{\xi_2\,\xi_1}\, p_C^{(\xi_1\, gg)}[t_{CA}]_{gg\,gg}|\psi_0\rangle\,.$$

The overlap in eqn (7.58) serves as a template for the others appearing in eqns (7.49) and (7.50), which can be written out directly:

$$\langle\uparrow_A\downarrow_C\uparrow_D\mid\uparrow_B\rangle = -e^{-i\varphi_{BA}+i\varphi_{DC}} \sum_{\xi_1\,\xi_2\,\xi_3\,\xi_4} F_A^{(gg\,\xi_4)} F_C^{(\xi_3\, gg)} F_D^{(gg\,\xi_2)} F_B^{(\xi_1\, gg)} \tag{7.59}$$
$$\times \langle\psi_0|p_A^{(gg\,\xi_4)}[-t_{CA}]_{\xi_4\,\xi_3}\, p_C^{(\xi_3\, gg)}[-t_{DC}]_{gg\,gg}\, p_D^{(gg\,\xi_2)}[t_{DB}]_{\xi_2\,\xi_1}\, p_B^{(\xi_1\, gg)}[t_{BA}]_{gg\,gg}|\psi_0\rangle\,,$$

$$\langle\uparrow_A\mid\uparrow_B\uparrow_C\downarrow_D\rangle = -e^{-i\varphi_{BA}+i\varphi_{DC}} \sum_{\xi_1\,\xi_2\,\xi_3\,\xi_4} F_A^{(gg\,\xi_1)} F_D^{(\xi_2\, ee)} F_C^{(ee\,\xi_3)} F_B^{(\xi_4\, gg)} \tag{7.60}$$
$$\times \langle\psi_0|p_A^{(gg\,\xi_1)}[-t_{DA}]_{\xi_1\,\xi_2}\, p_D^{(\xi_2\, ee)}[t_{DC}]_{ee\,ee}\, p_C^{(ee\,\xi_3)}[t_{CB}]_{\xi_3\,\xi_4}\, p_B^{(\xi_4\, gg)}[t_{BA}]_{gg\,gg}|\psi_0\rangle\,,$$

and

$$\langle\uparrow_A\uparrow_D\mid\uparrow_B\uparrow_C\rangle = e^{-i\varphi_{BA}+i\varphi_{DC}} \sum_{\xi_1\,\xi_2\,\xi_3\,\xi_4} F_A^{(gg\,\xi_1)} F_D^{(\xi_2\, ee)} F_C^{(ee\,\xi_3)} F_B^{(\xi_4\, gg)} \tag{7.61}$$
$$\times \langle\psi_0|p_A^{(gg\,\xi_1)}[-t_{DA}]_{\xi_1\,\xi_2}\, p_D^{(\xi_2\, ee)}[t_{DC}]_{ee\,ee}\, p_C^{(ee\,\xi_3)}[t_{CB}]_{\xi_3\,\xi_4}\, p_B^{(\xi_4\, gg)}[t_{BA}]_{gg\,gg}|\psi_0\rangle\,.$$

It is to be noted that $\langle\uparrow_A\uparrow_D\mid\uparrow_B\uparrow_C\rangle$, which describes excited-state absorption from the singly to the doubly excited electronic manifold, equals minus $\langle\uparrow_A\mid\uparrow_B\uparrow_C\downarrow_D\rangle$, which accounts for "bleaching" of the singly excited states in the same process. The degree to which these overlaps cancel in the overall WPI signal depends on the relative fluorescence quantum yield following double and single excitation.

Each of the overlaps (7.58)–(7.61) contains two periods of free molecular evolution in the singly excited manifold of the form $[t_{KL}]_{\xi\,\bar{\xi}}$ during which site-to-site electronic excitation transfer may not or must occur, according to whether ξ is or isn't equal to $\bar{\xi}$; these free molecular-evolution operators consist exclusively of terms of even or odd powers of J, respectively. Thus, an expansion of the signal through first order in J could be found by collecting terms of zeroth and first order in eqns (7.58)–(7.61). But the task is simplified when we consider the chosen dimer orientation and the polarization directions. For the A-pulse we have

$$F_A^{(gg\,eg)} = 0\,;$$
$$F_A^{(gg\,ge)} = m_b F_A \tag{7.62}$$

(see eqn (7.48)). While for $I = B, C, D$,

$$F_I^{(gg\,\xi)} = \delta_{\xi\,eg} m_a F_I\,;$$
$$F_I^{(eg\,\xi)} = \delta_{\xi\,gg} m_a F_I\,;$$
$$F_I^{(ge\,\xi)} = \delta_{\xi\,ee} m_a F_I\,;$$
$$F_I^{(ee\,\xi)} = \delta_{\xi\,ge} m_a F_I\,. \tag{7.63}$$

Under these circumstances, net transfer occurs during one particular evolution interval in each of the contributing overlaps. All of them vanish at zeroth order in J, and the sought-after first-order overlaps reduce to

$$\langle \uparrow_A \downarrow_B \uparrow_D \mid \uparrow_C \rangle = -e^{-i\varphi_{BA}+i\varphi_{DC}} m_a^3 m_b F_A F_B F_C F_D \tag{7.64}$$
$$\times \langle \psi_0 | p_A^{(gg\,ge)} [-t_{BA}]^{(1)}_{ge\,eg}\, p_B^{(eg\,gg)} [-t_{DB}]_{gg\,gg}\, p_D^{(gg\,eg)} [t_{DC}]^{(0)}_{eg\,eg}\, p_C^{(eg\,gg)} [t_{CA}]_{gg\,gg} | \psi_0 \rangle\,,$$

$$\langle \uparrow_A \downarrow_C \uparrow_D \mid \uparrow_B \rangle = -e^{-i\varphi_{BA}+i\varphi_{DC}} m_a^3 m_b F_A F_B F_C F_D \tag{7.65}$$
$$\times \langle \psi_0 | p_A^{(gg\,ge)} [-t_{CA}]^{(1)}_{ge\,eg}\, p_C^{(eg\,gg)} [-t_{DC}]_{gg\,gg}\, p_D^{(gg\,eg)} [t_{DB}]^{(0)}_{eg\,eg}\, p_B^{(eg\,gg)} [t_{BA}]_{gg\,gg} | \psi_0 \rangle\,,$$

$$\langle \uparrow_A \mid \uparrow_B \uparrow_C \downarrow_D \rangle = -e^{-i\varphi_{BA}+i\varphi_{DC}} m_a^3 m_b F_A F_B F_C F_D \tag{7.66}$$
$$\times \langle \psi_0 | p_A^{(gg\,ge)} [-t_{DA}]^{(0)}_{ge\,ge}\, p_D^{(ge\,ee)} [t_{DC}]_{ee\,ee}\, p_C^{(ee\,ge)} [t_{CB}]^{(1)}_{ge\,eg}\, p_B^{(eg\,gg)} [t_{BA}]_{gg\,gg} | \psi_0 \rangle\,,$$

and

$$\langle \uparrow_A \uparrow_D \mid \uparrow_B \uparrow_C \rangle = -\langle \uparrow_A \mid \uparrow_B \uparrow_C \downarrow_D \rangle\,. \tag{7.67}$$

Free-evolution operators of zeroth and first order in J within the singly excited manifold are marked with superscripts. More explicitly, $[t] \cong [t]^{(0)} + [t]^{(1)}$, where $[t]^{(0)} = \exp\{-iH^{(0)}t/\hbar\}$ (with $H^{(0)}$ being $T + H_{el}(Q)$ in which J is set to zero) and, by first-order time-dependent perturbation theory,

$$[t]^{(1)} = -\frac{iJ}{\hbar}\int_0^t d\tau\, [t-\tau]^{(0)} \big(|eg\rangle\langle ge| + |ge\rangle\langle eg|\big) [\tau]^{(0)}\,. \tag{7.68}$$

In the calculations shown here, the common envelope function for all four pulses is taken to be $f(t) = \exp\{-t^2/2\sigma^2\}$ with $\sigma = 0.12(2\pi/\omega) \equiv 0.12\tau_v$, about an eighth of a vibrational period. From eqn (7.47) the reduced pulse propagators,

$$p^{(\xi\bar{\xi})} = \int_{-\infty}^{\infty} \frac{d\tau}{\sigma} f(\tau) e^{\mp i\Omega\tau} [-\tau]^{(0)}_{\xi\xi} [\tau]^{(0)}_{\bar{\xi}\bar{\xi}} , \tag{7.69}$$

are seen to become proportional to Fourier components of the envelope evaluated at the offset between Ω (here set to $(\epsilon_{eg} + \epsilon_{ge})/2\hbar$) and the $\bar{\xi}$-to-ξ vibronic transition frequency.

In order to calculate the WPI signal from the excited-state populations, we have to specify the relative fluorescence quantum yield from the singly and doubly excited manifolds. We set $Q_{one} = 1$ and illustrate several limiting possibilities by taking $Q_{two} = 0, 1, \text{or } 2$. The first choice would be appropriate if some rapid, nonradiative process shuts off fluorescence from doubly excited dimers; the second would apply if the dimer undergoes rapid internal conversion to the singly excited manifold prior to radiative decay; and the last value would be applicable if simultaneously excited monomers within a dimer decay by emitting one photon each.

We have not yet stated a value for the bare electronic energy ϵ_{ee} seen in eqn (7.54). One could imagine that it is simply the sum of the individual excitation energies, ϵ_{eg} and ϵ_{ge}, and some of our calculations will make this assumption. It's also possible, however, that an "exciton shift" alters the excitation energy for one monomer when the other is already excited, perhaps lowering it due to stronger dispersion interactions between two excited monomers than between one excited and another unexcited species. We'll entertain the possible effect of an exciton shift by considering an alternative choice, $\epsilon_{ee} = \epsilon_{eg} + \epsilon_{ge} - \hbar\omega/2$. Even this small shift will be seen to influence non-negligibly the form of the calculated interferograms, due to the phase-sensitive nature of 2D WPI. A significantly stronger exciton shift might even move singly-to-doubly excited transitions outside the power spectra of finite-duration laser pulses, driving the relevant matrix elements of $p_C^{(ee\,ge)}$ and $p_D^{(ge\,ee)}$ to zero in eqns (7.66) and (7.67) and eliminating contributions from those overlaps. This consequence of a strong exciton shift would be similar to the cancelation between these two overlaps that occurs when $Q_{two} = Q_{one}$.

Shown below are the real part and the absolute value of each of the contributing overlaps and the 2D-WPI signal, calculated under the conditions just described, with all the time-evolution operators and pulse propagators represented as matrices in a two-dimensional harmonic oscillator basis. The waiting time is set to half a vibrational period ($t_{CB} = \tau_v/2$) to allow optically generated excited-state wave packets at most one pass through the $V_{eg} = V_{ge}$ intersection during this interval. The overlaps are multiplied by $\exp\{i\varphi_{BA} - i\varphi_{DC} + i\Omega(t_{DC} - t_{BA})\}$ in order to eliminate the controlled optical phase shifts and to alias the electronic-frequency oscillations. All of the overlaps and signals in this section, which should be dimensionless, are plotted in units of $J\tau_v/\hbar$ times $m_a^3 m_b F_A F_B F_C F_D$.

Figure 7.8 plots $\langle \uparrow_A \downarrow_B \uparrow_D \mid \uparrow_C \rangle$ and $\langle \uparrow_A \downarrow_C \uparrow_D \mid \uparrow_B \rangle$, which appear in $P_{one}^{(d)}$. Since the three-pulse bra in each of these overlaps does not access the doubly excited state, they would be impervious to an exciton shift. In Fig. 7.9 is shown $\langle \uparrow_A \mid \uparrow_B \uparrow_C \downarrow_D \rangle$, also appearing in $P_{one}^{(d)}$, whose three-pulse ket involves *ee*-state wave-packet dynamics between t_C and t_D. This overlap will be recalculated below with an exciton shift included. It is not necessary to make a separate plot for the overlap $\langle \uparrow_A \uparrow_D \mid \uparrow_B \uparrow_C \rangle$ determining $P_{two}^{(d)}$, as it equals minus $\langle \uparrow_A \mid \uparrow_B \uparrow_C \downarrow_D \rangle$ without or with an exciton shift.

For the model dimer under study, it is possible to understand many key features of the delay dependence of the overlaps shown in Figs 7.8 and 7.9 in terms of the underlying energy-transfer and nuclear wave-packet dynamics. The vanishing of $\langle \uparrow_A \downarrow_B \uparrow_D \mid \uparrow_C \rangle$ for $t_{BA} = 2n\tau_v$ is a striking consequence of the half-quantum offset between ϵ_{eg} and ϵ_{ge}. We see from eqn (7.64) that first-order energy transfer must occur during t_{BA} for this overlap to be nonzero. When $t_{BA} = 2n\tau_v$, this EET event could take place during any $2\tau_v$-long interval, so we can write

$$[-2n\tau_v]_{ge\,eg}^{(1)} = [-2\tau_v]_{ge\,eg}^{(1)} [-2(n-1)\tau_v]_{eg\,eg}^{(0)} + [-2\tau_v]_{ge\,ge}^{(0)} [-2\tau_v]_{ge\,eg}^{(1)} [-2(n-2)\tau_v]_{eg\,eg}^{(0)} + \cdots + [-2(n-1)\tau_v]_{ge\,ge}^{(0)} [-2\tau_v]_{ge\,eg}^{(1)} . \tag{7.70}$$

We can break up the integral

$$\begin{aligned}
[-2\tau_v]_{ge\,eg}^{(1)} &= -\frac{iJ}{\hbar}\int_0^{-2\tau_v} d\tau\, [-2\tau_v - \tau]_{ge\,ge}^{(0)} [\tau]_{eg\,eg}^{(0)} \\
&= -\frac{iJ}{\hbar}\int_0^{-\tau_v} d\tau\, [-2\tau_v - \tau]_{ge\,ge}^{(0)} [\tau]_{eg\,eg}^{(0)} \\
&\quad - \frac{iJ}{\hbar}\int_{-\tau_v}^{-2\tau_v} d\tau\, [-2\tau_v - \tau]_{ge\,ge}^{(0)} [\tau]_{eg\,eg}^{(0)} \\
&= -\frac{iJ}{\hbar}\int_0^{-\tau_v} d\tau\, [-2\tau_v - \tau]_{ge\,ge}^{(0)} [\tau]_{eg\,eg}^{(0)} \\
&\quad - \frac{iJ}{\hbar} e^{i\tau_v(\epsilon_{eg} - \epsilon_{ge})} \int_0^{-\tau_v} d\bar{\tau}\, [-2\tau_v - \bar{\tau}]_{ge\,ge}^{(0)} [\bar{\tau}]_{eg\,eg}^{(0)} = 0\,;
\end{aligned} \tag{7.71}$$

the nuclear wave packet generated in the *eg*-state by EET during the second half of each $2\tau_v$-interval is opposite in sign from that formed during the first half, giving rise to complete destructive interference and no net amplitude transfer. The first-order evolution operator in eqn (7.70) therefore vanishes entirely.

Devise an argument explaining why $\langle \uparrow_A \downarrow_C \uparrow_D \mid \uparrow_B \rangle$ vanishes at $t_{BA} = (2n - \frac{1}{2})\tau_v$.

A condition for $\langle \uparrow_A \downarrow_B \uparrow_D \mid \uparrow_C \rangle$ to have a sizable absolute value is that the overlapped one- and three-pulse wave packets reside in similar regions of phase space. That is to say, the expectation values of their coordinate and momentum must coincide, for both *a* and *b* modes. Schematic diagrams for both modes of both wave packets are sketched in Fig. 7.10. The *A* pulse excites the dimer to the *ge* state, where *b*-mode motion ensues. The local splitting between the two site-states becomes larger than its Franck–Condon value of $\hbar\omega/2$ at positive q_b; so the most likely elapsed times before an energy-transfer transition to *eg*, denoted by t_{JA}, will be integer multiples of τ_v. The condition for

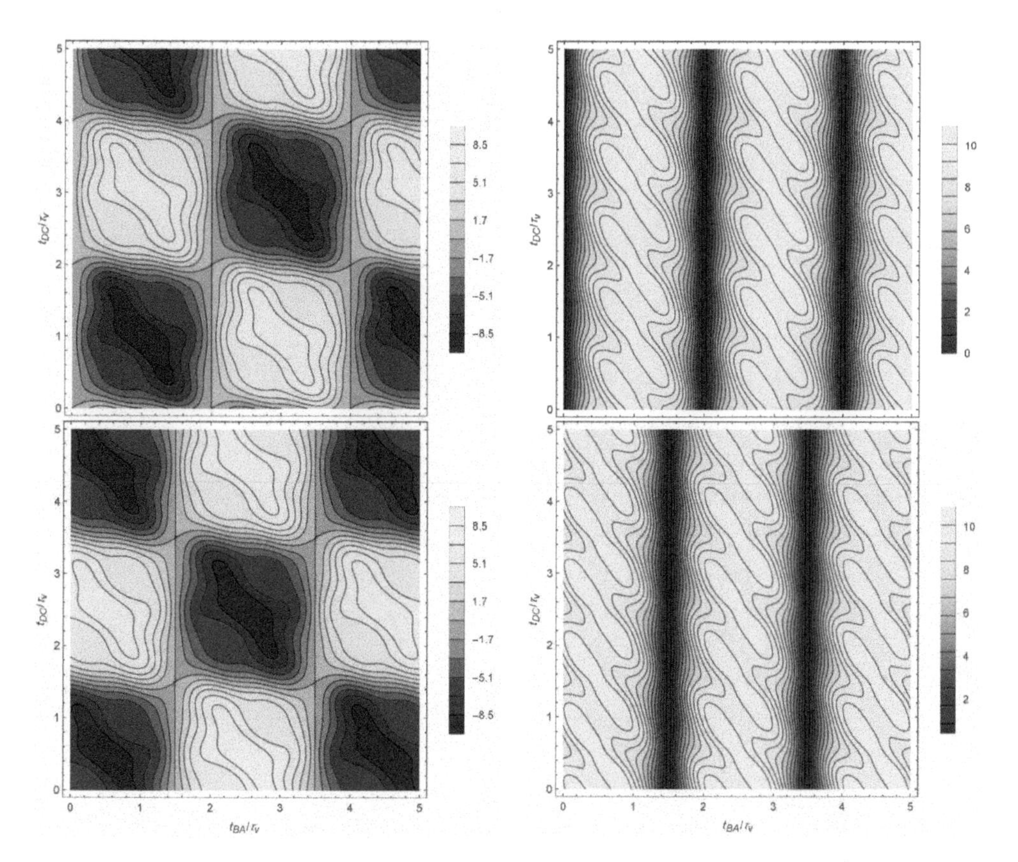

Fig. 7.8 Two upper panels are the real part (left) and absolute value (right) of $\langle \uparrow_A \downarrow_B \uparrow_D | \uparrow_C \rangle$; the former is phase-referenced and frequency aliased as described in the text. Intrapulse-pair delays are in vibrational periods $\tau_v = 2\pi/\omega$, and the waiting time is fixed at $t_{CB} = \tau_v/2$. Lower panels are for $\langle \uparrow_A \downarrow_C \uparrow_D | \uparrow_B \rangle$. Neither of these overlaps would be affected by an exciton shift in the singly-to-doubly excited transition energy.

a-mode coincidence can then be written

$$\begin{aligned} \omega d(1 - e^{-i\omega t_{DC}}) &= \omega d(1 - e^{-i\omega t_{BJ}})e^{-i\omega t_{DB}} \\ &\approx \omega d(1 - e^{-i\omega t_{BA}})e^{-i\omega t_{DC} - i\omega \frac{\tau_v}{2}} , \end{aligned} \tag{7.72}$$

which reduces to $t_{DC} + t_{BA} = m\tau_v$ and rationalizes the slanted form of the peaks in the upper right panel of Fig. 7.8. The b-mode coincidence requirement $0 = \omega d(1 - \exp\{-i\omega t_{JA}\}) \exp\{-i\omega t_{DJ}\}$ simply reinforces the condition $t_{JA} \approx n\tau_v$.

Carry out a similar phase-space coincidence analysis for $\langle \uparrow_A \downarrow_C \uparrow_D | \uparrow_B \rangle$.

The exciton shift affects the two contributing overlaps which involve wave-packet motion in the doubly excited electronic state. $\langle \uparrow_A | \uparrow_B \uparrow_C \downarrow_D \rangle$ with the chosen exciton shift (see page 111) is illustrated in Fig. 7.11. $\langle \uparrow_A \uparrow_D | \uparrow_B \uparrow_C \rangle$ has the opposite sign.

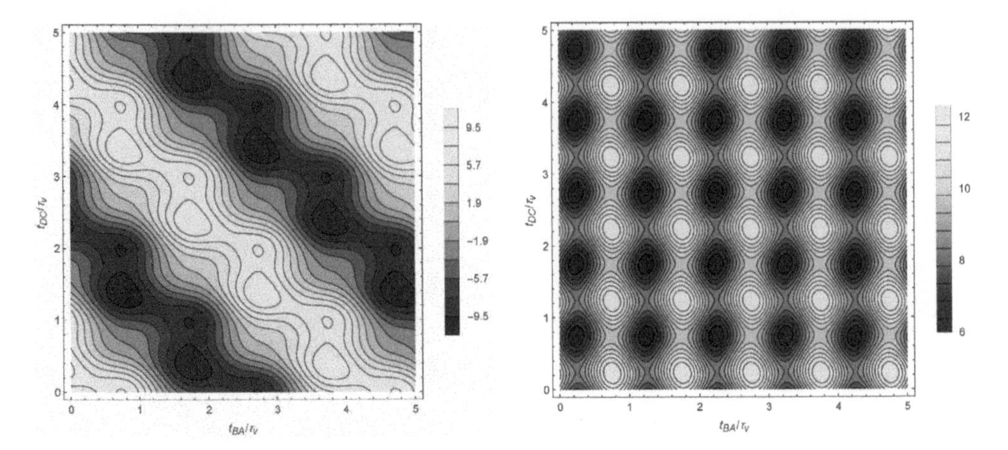

Fig. 7.9 Left panel gives $\mathrm{Re}\{\langle\uparrow_A \mid \uparrow_B\uparrow_C\downarrow_D\rangle\}$ (or $-\mathrm{Re}\{\langle\uparrow_A\uparrow_D \mid \uparrow_B\uparrow_C\rangle\}$) for the model dimer without an exciton shift, and right panel shows its absolute value.

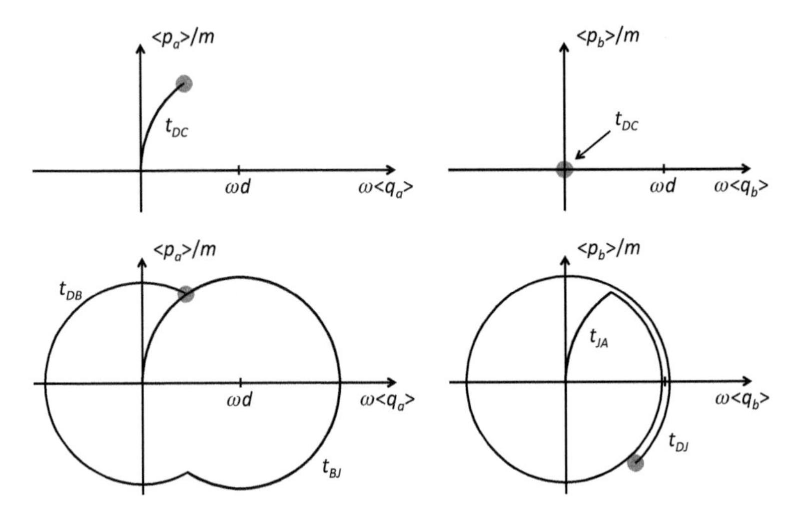

Fig. 7.10 The upper two panels show approximate phase-space paths for $\mid\uparrow_C\rangle$ while the lower two show those for $\mid\uparrow_A\downarrow_B\uparrow_D\rangle$, in the case $t_{CB} = \tau_v/2$.

The complex-valued overlap itself changes drastically between Fig. 7.9 and Fig. 7.11, illustrating the sensitivity of the overlaps determining the 2D-WPI signal to small changes in the relative phase of the interfering wave packets. The absolute value of the overlap changes only slightly as a result of this small shift, reflecting minute changes in the location of the dimer's singly-to-doubly excited vibronic transition energies within the power spectrum of the pulses.

The bra and ket trajectories whose final-point coincidence determines the delay combinations of maximal overlap visible in the right panels of Figs 7.9 and 7.11 are drawn schematically in Fig. 7.12. The requirement for agreement between the two a-mode ending points can be written

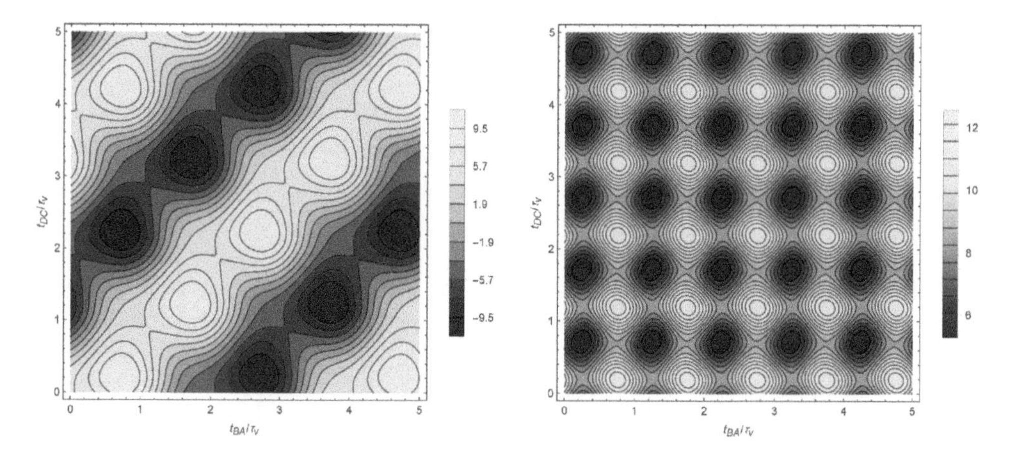

Fig. 7.11 Left panel shows $\mathrm{Re}\{\langle\uparrow_A \mid \uparrow_B\uparrow_C\downarrow_D\rangle\}$ (or $-\mathrm{Re}\{\langle\uparrow_A\uparrow_D \mid \uparrow_B\uparrow_C\rangle\}$) for the model dimer *with* an exciton shift, and right panel plots its absolute value.

$$0 = \omega d\big[\big(1 - e^{-i\omega t_{JB}}\big)e^{-i\omega(t_{CB}-t_{JB})} - 1\big]e^{-i\omega t_{DC}} + \omega d\,, \tag{7.73}$$

where $t_{CB} = \tau_v/2$ and t_{JB} estimates the delay between B-pulse arrival and amplitude transfer by EET from eg to ge. This requirement reduces to $t_{DC} = k\tau_v + t_{JB}$. The corresponding condition for the b-mode is

$$\omega d\big(1 - e^{-i\omega t_{DA}}\big) = \omega d\big(1 - e^{-i\omega t_{DJ}}\big)\,, \tag{7.74}$$

or $t_{BA} = l\tau_v - t_{JB}$. Since $t_{JB} \approx \tau_v/4$ in the present situation, the coincidence requirements predict peaks at $(t_{BA}, t_{DC}) \approx (l - \frac{1}{4}, k + \frac{1}{4})\tau_v$, just as seen in the absolute-value plots.

> Draw phase-space trajectories for the bra and ket of $\langle\uparrow_A\uparrow_D \mid \uparrow_B\uparrow_C\rangle$ and demonstrate that agreement between their end points leads to the same (t_{BA}, t_{DC}) values as $\langle\uparrow_A \mid \uparrow_B\uparrow_C\downarrow_D\rangle$.

Despite the close similarity between the right-hand panels of Figs 7.9 and 7.11, the phase dependence exhibited on the left is quite different. The residual electronic phase factor of the aliased overlap is

$$\begin{aligned} e^{i\Omega(t_{DC}-t_{BA})}\langle\uparrow_A \mid \uparrow_B\uparrow_C\downarrow_D\rangle \sim \exp\Bigg\{ & \frac{i}{\hbar}\left(\epsilon_{ge} + \frac{\Delta\epsilon}{2}\right)(t_{DC} - t_{BA}) \\ & + \frac{i}{\hbar}\epsilon_{ge}\left(t_{DC} + \frac{\tau_v}{2} + t_{BA}\right) - \frac{i}{\hbar}(2\epsilon_{eg} + \Delta\epsilon + \delta\epsilon)t_{DC} \\ & + \frac{i}{\hbar}\epsilon_{ge}\left(t_{JB} - \frac{\tau_v}{2}\right) - \frac{i}{\hbar}(\epsilon_{eg} + \Delta\epsilon)t_{JB}\Bigg\} \\ = \exp\Bigg\{ & -\frac{i}{\hbar}\Delta\epsilon\left(\frac{t_{DC} + t_{BA}}{2} + t_{JB}\right) - \frac{i}{\hbar}\delta\epsilon\, t_{DC}\Bigg\} \end{aligned} \tag{7.75}$$

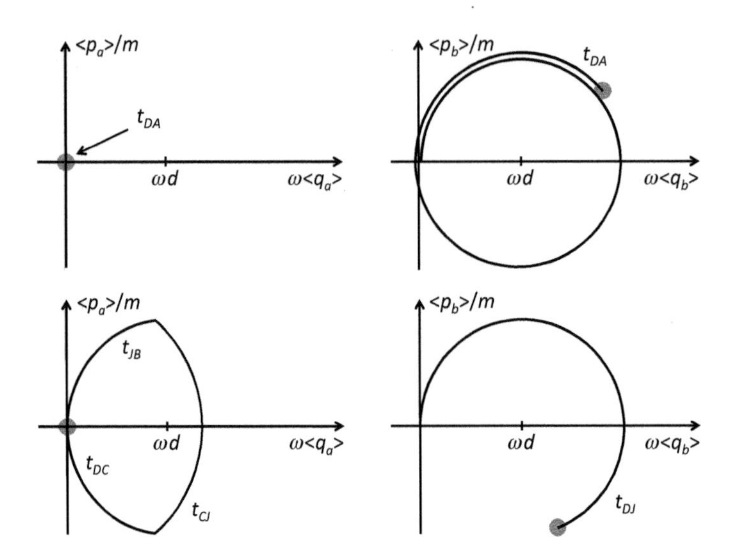

Fig. 7.12 The top (bottom) panels show momentum-versus-position expectation-value trajectories for the bra (ket) of $\langle\uparrow_A \mid \uparrow_B\uparrow_C\downarrow_D\rangle$. Coincidence between the endpoints of these phase-space paths is a prerequisite for this overlap to be large.

(see page 111), where $\Delta\epsilon = \epsilon_{eg} - \epsilon_{ge} = \hbar\omega/2$ is the site-energy offset and $\delta\epsilon = 0, -\hbar\omega/2$ is the exciton shift. In the absence of the latter,

$$e^{i\Omega(t_{DC}-t_{BA})}\langle\uparrow_A \mid \uparrow_B\uparrow_C\downarrow_D\rangle \sim \exp\left\{-i\omega\left(\frac{t_{DC}+t_{BA}}{4}+\frac{t_{JB}}{2}\right)\right\}, \tag{7.76}$$

while in its presence,

$$e^{i\Omega(t_{DC}-t_{BA})}\langle\uparrow_A \mid \uparrow_B\uparrow_C\downarrow_D\rangle \sim \exp\left\{i\omega\left(\frac{t_{DC}-t_{BA}}{4}-\frac{t_{JB}}{2}\right)\right\}. \tag{7.77}$$

Equations (7.76) and (7.77) account for the constant phase of the overlap along lines of constant $t_{DC}+t_{BA}$ and $t_{DC}-t_{BA}$ seen in Figs 7.9 and 7.11, respectively.

7.3.2 Signals

While the delay dependence of the individual overlaps is relatively easy to analyze in terms of wave-packet dynamics, the 2D-WPI signal (eqn (7.37)) comprises a quantum-yield-weighted sum of the several contributing overlaps. Under the assumptions described on page 102, the difference-phased whoopee signal from the model dimer becomes $P^{(d)} = 2\mathrm{Re}\{\xi_d e^{-i\varphi_{BA}+i\varphi_{DC}}\}$, where

$$\begin{aligned}\xi_d e^{-i\varphi_{BA}+i\varphi_{DC}} = Q_{one}\{&\langle\uparrow_A\downarrow_B\uparrow_D \mid \uparrow_C\rangle + \langle\uparrow_A\downarrow_C\uparrow_D \mid \uparrow_B\rangle \\ &+ \langle\uparrow_A \mid \uparrow_B\uparrow_C\downarrow_D\rangle\} + Q_{two}\langle\uparrow_A\uparrow_D \mid \uparrow_B\uparrow_C\rangle\,.\end{aligned} \tag{7.78}$$

Since one of the equal-and-opposite overlaps involving access to the doubly excited state contributes to $P^{(d)}_{one}$ and the other determines $P^{(d)}_{two}$, their degree of cancellation will depend on the relative quantum yield from populations in the singly- and doubly-excited manifolds. We shall examine calculated signals for $Q_{one} = 1$ and $Q_{two} = 0, 1,$ and 2. As explained in Chapter 6, the complex-valued interferogram ξ_d can be isolated by combining signals $P^{(d)}$ with various choices of intrapulse-pair optical phase shifts.

Figure 7.13 presents the aliased 2D interferogram $\xi_d e^{i\Omega(t_{DC}-t_{BA})}$ in the case $Q_{two} = 0$. In the top (bottom) row are the real part and absolute value of the WPI signal without (with) an exciton shift. Although $\langle \uparrow_A \uparrow_D \mid \uparrow_B \uparrow_C \rangle$ fails to contribute in this case, the interferogram remains sensitive to the exciton shift through $\langle \uparrow_A \mid \uparrow_B \uparrow_C \downarrow_D \rangle$; both the real part and the absolute value differ markedly in the two cases.

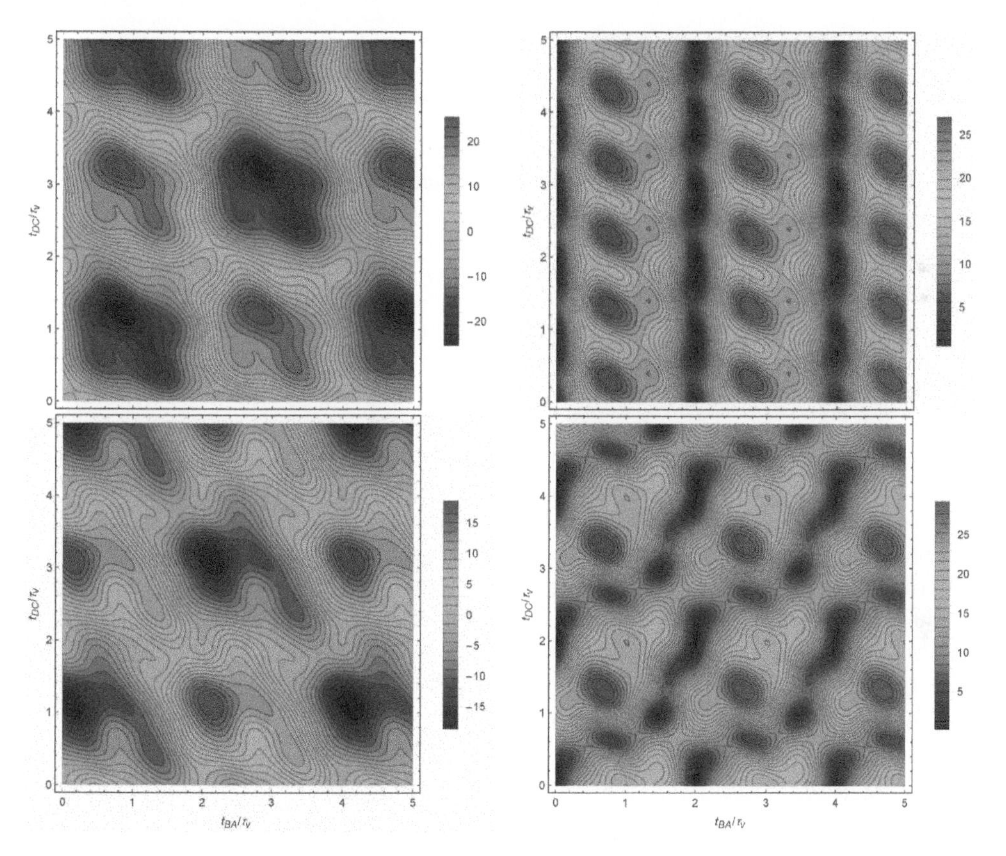

Fig. 7.13 The top two panels exhibit $\mathrm{Re}\{\xi_d e^{i\Omega(t_{DC}-t_{BA})}\}$ (left) and $\mathrm{Abs}\{\xi_d e^{i\Omega(t_{DC}-t_{BA})}\}$ (right) for the weakly coupled EET dimer with $Q_{two} = 0$ in the absence of an exciton shift. Their forms in the presence of a small, negative shift are shown in the bottom row.

If $Q_{two} = Q_{one} = 1$, the case of equal fluorescence yields from the singly and doubly excited manifolds, then $\langle \uparrow_A \mid \uparrow_B \uparrow_C \downarrow_D \rangle$ and $\langle \uparrow_A \uparrow_D \mid \uparrow_B \uparrow_C \rangle$ cancel each other exactly; the interference signal becomes independent of any exciton shift. The 2D-WPI signal, illustrated in Fig. 7.14, is now determined by the two remaining overlaps contributing to $P^{(d)}_{one}$.

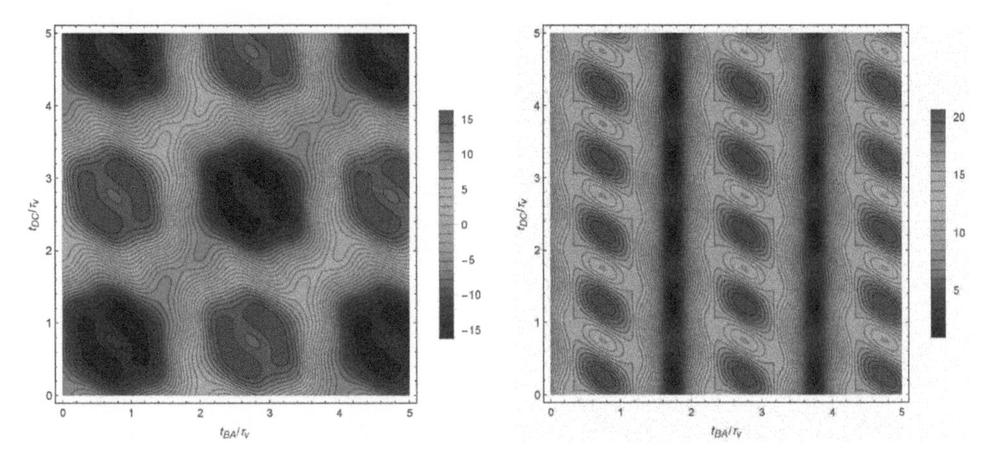

Fig. 7.14 Real part and absolute value of the aliased WPI signal $\xi_d e^{i\Omega(t_{DC}-t_{BA})}$ for $Q_{two} = Q_{one} = 1$, which are not affected by an exciton shift.

Sensitivity to the exciton shift returns when $Q_{two} = 2$, meaning that doubly excited states are twice as productive of fluorescent photons as singly excited states. Here, the contribution from $\langle \uparrow_A \uparrow_D \mid \uparrow_B \uparrow_C \rangle$ outweighs that from $\langle \uparrow_A \mid \uparrow_B \uparrow_C \downarrow_D \rangle$. The EET dimer's signal for this yield ratio is shown in Fig. 7.15.

The calculations presented in this section illustrate the physical information content and dynamical interpretation of two-dimensional wave-packet interferometry signals from an energy-transfer dimer under a specific set of molecular and experimental parameters. Many elaborations and generalizations remain to be investigated. While the present illustrative calculations yield interferograms of undiminished signal intensity with increasing t_{BA} and t_{DC}, more realistic simulations including (perhaps weak) electronic-nuclear coupling for a large number of intra- and intermolecular modes would exhibit "optical dephasing." As a result, increasing intrapulse-pair delays would be accompanied by decreasing signal size, akin to that seen in Fig. 6.13 of Chapter 6, as electronic-nuclear entanglement destroys the electronic coherence required for a nonvanishing 2D-WPI signal.

7.3.3 Whoopee signal in a strong-J case

For comparison with the weak EET-coupling case, we now investigate the 2D-WPI signal from our model system when $2J > m\omega^2 d^2$, choosing $J = (\sqrt{7}/2)\hbar\omega$ and, as before, setting $d = \sqrt{\hbar/2m\omega}$. We keep the same site-dipole orientations and the same sequence of laser polarizations as previously, but set $\Delta\epsilon = 0$ to eliminate the difference in site energies. We further assume equal fluorescence quantum yield from the singly- and doubly-excited states ($Q_{two} = Q_{one}$), so the contributions of excited-state absorption $\langle \uparrow_A \uparrow_D \mid \uparrow_B \uparrow_C \rangle$ and excited-state bleach $\langle \uparrow_A \mid \uparrow_B \uparrow_C \downarrow_D \rangle$ to $P^{(d)}$ cancel—see eqns (7.50) and (7.49), respectively—and the possible presence of an exciton shift becomes immaterial. The difference-phased WPI signal now reduces to

$$P^{(d)} = 2Q_{one}\mathrm{Re}\{\langle \uparrow_A \downarrow_B \uparrow_D \mid \uparrow_C \rangle + \langle \uparrow_A \downarrow_C \uparrow_D \mid \uparrow_B \rangle\}\,. \tag{7.79}$$

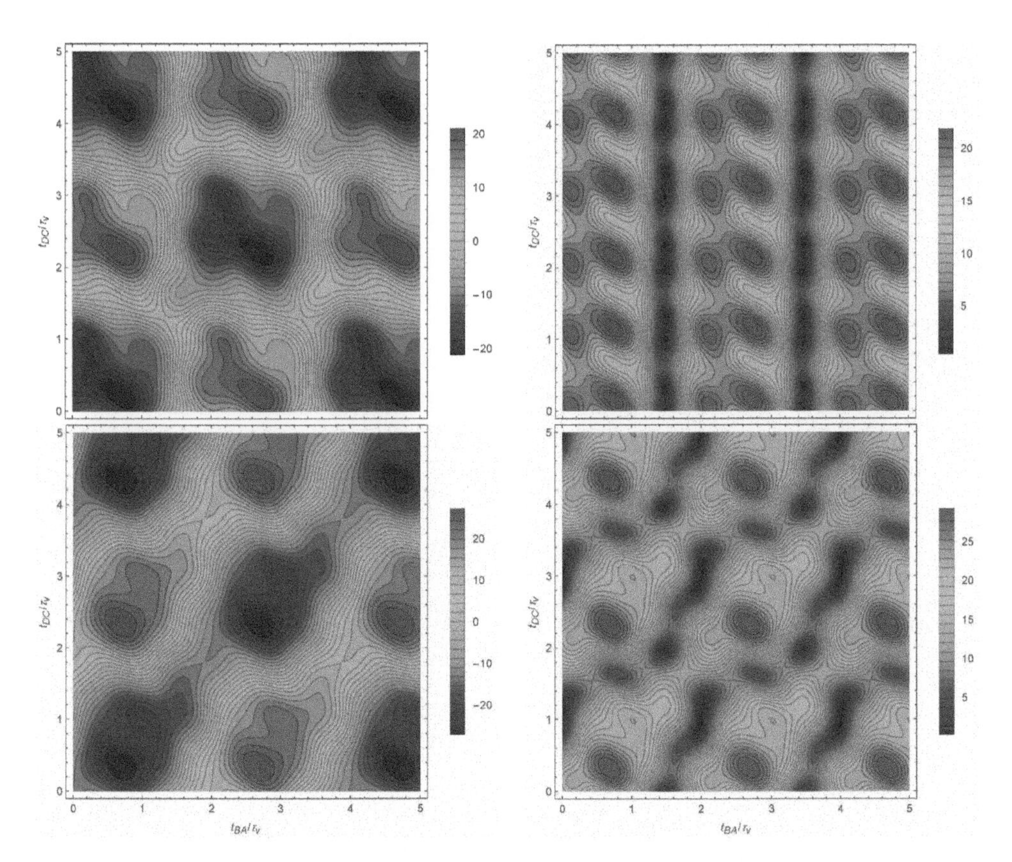

Fig. 7.15 Top two panels show real part (left) and absolute value (right) of small-J dimer's WPI signal when $Q_{two} = 2Q_{one}$ and there is no exciton shift, while bottom two include a small down-shift for singly-to-doubly excited transitions.

In this strong energy-transfer case, the exciton basis is the practical choice, and the electronic Hamiltonian of eqn (7.24) becomes

$$\begin{aligned} H_{el}(q_a, q_b) = {} & |0\rangle\langle 0| E_0(q_a, q_b) + |\bar{1}\rangle\langle\bar{1}| \big(L(q_a, q_b) + J\big) + |1\rangle\langle 1| \big(L(q_a, q_b) - J\big) \\ & - \big(|\bar{1}\rangle\langle 1| + |1\rangle\langle\bar{1}|\big) K(q_a, q_b) + |2\rangle\langle 2| E_2(q_a, q_b) \,, \end{aligned} \tag{7.80}$$

in which

$$E_0(q_a, q_b) = \tfrac{m\omega^2}{2}(q_a^2 + q_b^2) \,, \tag{7.81}$$

$$L(q_a, q_b) = \epsilon + \tfrac{m\omega^2}{2}\left[(q_a - \tfrac{d}{2})^2 + (q_b - \tfrac{d}{2})^2 + \tfrac{d^2}{2}\right] , \tag{7.82}$$

$$K(q_a, q_b) = \tfrac{m\omega^2}{2} d(q_b - q_a) \,, \tag{7.83}$$

and (although we don't need it here)

$$E_2(q_a, q_b) = 2\epsilon + \delta\epsilon + \tfrac{m\omega^2}{2}\left[(q_a - d)^2 + (q_b - d)^2\right] . \tag{7.84}$$

The eigenstates of $H_{el}(0, 0)$, constituting the exciton basis, are given by

$$|0\rangle = |gg\rangle\,, \tag{7.85}$$

$$|\bar{1}\rangle = \tfrac{1}{\sqrt{2}}(|ge\rangle + |eg\rangle)\,, \tag{7.86}$$

$$|1\rangle = \tfrac{1}{\sqrt{2}}(|ge\rangle - |eg\rangle)\,, \tag{7.87}$$

and

$$|2\rangle = |ee\rangle\,; \tag{7.88}$$

the one-exciton states $|\bar{1}\rangle$ and $|1\rangle$ are obtained from eqns (7.16) and (7.17), respectively, by setting the rotation angle to $\theta(0,0) = \pi/2$.

Since the interexciton coupling term (eqn (7.83)) involves the difference $q_b - q_a$, it is helpful to switch to the alternative nuclear coordinates,

$$Q = \tfrac{1}{\sqrt{2}}(q_b + q_a) \quad \text{and} \quad q = \tfrac{1}{\sqrt{2}}(q_b - q_a)\,, \tag{7.89}$$

and momenta,

$$P = \tfrac{1}{\sqrt{2}}(p_b + p_a) \quad \text{and} \quad p = \tfrac{1}{\sqrt{2}}(p_b - p_a)\,. \tag{7.90}$$

In terms of these, the full dimer Hamiltonian is $H = H^{(0)} + H^{(1)}$, where $H^{(0)} = T + H_{el}^{(0)}(Q,q)$, with $T = \frac{1}{2m}(P^2 + p^2)$ and

$$H_{el}^{(0)}(Q,q) = |0\rangle\langle 0|E_0(Q,q) + |\bar{1}\rangle\langle\bar{1}|\big(L(Q,q) + J\big) \\ + |1\rangle\langle 1|\big(L(Q,q) - J\big) + |2\rangle\langle 2|E_2(Q,q)\,. \tag{7.91}$$

Here,

$$E_0(Q,q) = \tfrac{m\omega^2}{2}(Q^2 + q^2)\,, \tag{7.92}$$

$$L(Q,q) = \epsilon + \tfrac{m\omega^2}{2}\big[(Q - \tfrac{d}{\sqrt{2}})^2 + q^2 + \tfrac{d^2}{2}\big]\,, \tag{7.93}$$

and

$$E_2(Q,q) = 2\epsilon + \delta\epsilon + \tfrac{m\omega^2}{2}\big[(Q - \sqrt{2}d)^2 + q^2\big]\,. \tag{7.94}$$

Interexciton coupling is effected by the perturbation,

$$H^{(1)} = -\big(|\bar{1}\rangle\langle 1| + |1\rangle\langle\bar{1}|\big)K(q)\,, \tag{7.95}$$

where

$$K(q) = \tfrac{m\omega^2}{\sqrt{2}}d\,q\,. \tag{7.96}$$

Figure 7.16 illustrates the electronic dynamics at play in the two contributing overlaps with a y-polarized A pulse and x-polarized B, C, and D pulses. The y-polarized A pulse prepares an initial electronic superposition roughly equal to $|ge\rangle = (|\bar{1}\rangle + |1\rangle)/\sqrt{2}$, which sets the stage for coherent electronic excitation transfer. Both overlaps are of odd order in J, and the bra and ket in both may each involve multiple cycles of back-and-forth energy transfer between the two chromophores. But these sketches are more approximate than those of the weak-J case (Figs 7.5 and 7.6), because the pulse duration may no longer strictly beat the shortened energy-transfer timescale $h/2J$. Although it is now assumed that the influence of interexciton coupling

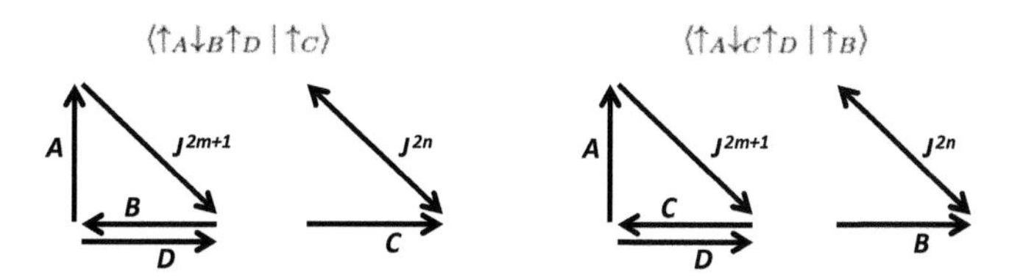

Fig. 7.16 Pulse-driven and energy transfer-mediated electronic transitions contributing to $P^{(d)}$ under $A_y B_x C_x D_x$ polarization in the strong-J case.

can be neglected in evaluating the reduced pulse propagators, the effects of excitation transfer must be included (see below eqn (7.47)).

Within a qualitative description that regards intrapulse energy transfer as slight, non-negligible values of $\langle \uparrow_A \downarrow_B \uparrow_D \mid \uparrow_C \rangle$, the overlap responsible for ground-state bleach and impulsive Raman excitation by the A-B pulse-pair, are expected for A-B delays equal to odd numbers of energy-transfer half-periods,

$$t_{BA} = (m + \tfrac{1}{2})\tfrac{h}{2J} \,, \tag{7.97}$$

and full-period C-D delays,

$$t_{DC} = n\tfrac{h}{2J} \,. \tag{7.98}$$

On the other hand, sizable stimulated-emission overlaps $\langle \uparrow_A \downarrow_C \uparrow_D \mid \uparrow_B \rangle$ can be anticipated for

$$t_{CB} + t_{BA} = (\bar{m} + \tfrac{1}{2})\tfrac{h}{2J} \,, \tag{7.99}$$

and

$$t_{DC} + t_{CB} = \bar{n}\tfrac{h}{2J} \,. \tag{7.100}$$

We see, in particular, that for a waiting time t_{CB} set to an integer multiple (odd half-integer multiple) of $h/2J$, the temporal locations of the ground-state-bleach/impulsive-Raman peaks should coincide with (be interleaved with) those of stimulated emission.

As suggested above, it's reasonable in this strong-J situation to suppose that the pulse durations σ are very short only with respect to the timescale $h/m\omega^2 d^2$ of interexciton coupling and perhaps only slightly shorter than $h/2J$. With the choice $\sigma = 0.06\tau_v$ and the selected values of d and J, we find $\sigma(m\omega^2 d^2/h) = 0.03$ and $\sigma(2J/h) = 0.16$.

Next, we work out explicit expressions for the two contributing overlaps. Using forms analogous to eqns (7.56) and (7.57), but in the exciton basis, we obtain

$$\langle \uparrow_A \downarrow_B \uparrow_D \mid \uparrow_C \rangle = -e^{-i\varphi_{BA} + i\varphi_{DC}} \sum_{\xi_1\, \xi_2\, \xi_3\, \xi_4 = \bar{1},1} F_A^{(0\xi_1)} F_B^{(\xi_2 0)} F_D^{(0\xi_3)} F_C^{(\xi_4 0)} \tag{7.101}$$
$$\times \langle \psi_0 | p_A^{(0\xi_1)}[-t_{BA}]_{\xi_1\xi_2} p_B^{(\xi_2 0)} [-t_{DB}]_{00} p_D^{(0\xi_3)} [t_{DC}]_{\xi_3\xi_4} p_C^{(\xi_4 0)} [t_{CA}]_{00} | \psi_0 \rangle \,.$$

which can be compared to eqn (7.58). We shall develop a treatment of the free-evolution operator in the one-exciton manifold that is perturbative in the interexciton coupling of eqn (7.95). Anticipating the fact that the zeroth-order contribution maintains the electronic state, we'll have

$$[t]^{(0)}_{\bar{\xi}\xi} = \delta_{\bar{\xi}\xi}[t]^{(0)}_{\xi\xi} \,. \tag{7.102}$$

As second-order evolution restores the electronic state, we also have

$$[t]^{(2)}_{\bar{\xi}\xi} = \delta_{\bar{\xi}\xi}[t]^{(2)}_{\xi\xi} \,, \tag{7.103}$$

while at first order, a transition occurs within the one-exciton manifold, so that

$$[t]^{(1)}_{\bar{\xi}\xi} = \delta_{\bar{\xi}\xi'}[t]^{(1)}_{\xi'\xi} \,, \tag{7.104}$$

where we have introduced the notation $\bar{1}' \equiv 1$ and $1' \equiv \bar{1}$.

At each order in the perturbation, two of the sums over exciton indices in eqn (7.101) reduce to a single term each, whence

$$\begin{aligned}\langle \uparrow_A \downarrow_B \uparrow_D \mid \uparrow_C \rangle^{(0)} &= -e^{-i\varphi_{BA}+i\varphi_{DC}} \sum_{\xi_1\,\xi_3} F_A^{(0\xi_1)} F_B^{(\xi_1 0)} F_D^{(0\xi_3)} F_C^{(\xi_3 0)} \\ &\times \langle \psi_0 | p_A^{(0\xi_1)} [-t_{BA}]^{(0)}_{\xi_1\xi_1} p_B^{(\xi_1 0)} [-t_{DB}]_{00} p_D^{(0\xi_3)} [t_{DC}]^{(0)}_{\xi_3\xi_3} p_C^{(\xi_3 0)} [t_{CA}]_{00} | \psi_0 \rangle \,,\end{aligned} \tag{7.105}$$

for the zeroth-order overlap,

$$\begin{aligned}\langle \uparrow_A \downarrow_B \uparrow_D \mid \uparrow_C \rangle^{(1)} &= -e^{-i\varphi_{BA}+i\varphi_{DC}} \sum_{\xi_1\,\xi_3} \Big\{ F_A^{(0\xi_1)} F_B^{(\xi_1' 0)} F_D^{(0\xi_3)} F_C^{(\xi_3 0)} \\ &\times \langle \psi_0 | p_A^{(0\xi_1)} [-t_{BA}]^{(1)}_{\xi_1\xi_1'} p_B^{(\xi_1' 0)} [-t_{DB}]_{00} p_D^{(0\xi_3)} [t_{DC}]^{(0)}_{\xi_3\xi_3} p_C^{(\xi_3 0)} [t_{CA}]_{00} | \psi_0 \rangle + F_A^{(0\xi_1)} F_B^{(\xi_1 0)} \\ &\times F_D^{(0\xi_3)} F_C^{(\xi_3' 0)} \langle \psi_0 | p_A^{(0\xi_1)} [-t_{BA}]^{(0)}_{\xi_1\xi_1} p_B^{(\xi_1 0)} [-t_{DB}]_{00} p_D^{(0\xi_3)} [t_{DC}]^{(1)}_{\xi_3\xi_3'} p_C^{(\xi_3' 0)} [t_{CA}]_{00} | \psi_0 \rangle \Big\} \,,\end{aligned} \tag{7.106}$$

for the first-order contribution, and

$$\begin{aligned}\langle \uparrow_A \downarrow_B \uparrow_D \mid \uparrow_C \rangle^{(2)} &= -e^{-i\varphi_{BA}+i\varphi_{DC}} \sum_{\xi_1\,\xi_3} \Big\{ F_A^{(0\xi_1)} F_B^{(\xi_1 0)} F_D^{(0\xi_3)} F_C^{(\xi_3 0)} \langle \psi_0 | p_A^{(0\xi_1)} \\ &\times \Big([-t_{BA}]^{(2)}_{\xi_1\xi_1} p_B^{(\xi_1 0)} [-t_{DB}]_{00} p_D^{(0\xi_3)} [t_{DC}]^{(0)}_{\xi_3\xi_3} + [-t_{BA}]^{(0)}_{\xi_1\xi_1} p_B^{(\xi_1 0)} [-t_{DB}]_{00} p_D^{(0\xi_3)} [t_{DC}]^{(2)}_{\xi_3\xi_3} \Big) \\ &\times p_C^{(\xi_3 0)} [t_{CA}]_{00} | \psi_0 \rangle + F_A^{(0\xi_1)} F_B^{(\xi_1' 0)} F_D^{(0\xi_3)} F_C^{(\xi_3' 0)} \langle \psi_0 | p_A^{(0\xi_1)} [-t_{BA}]^{(1)}_{\xi_1\xi_1'} p_B^{(\xi_1' 0)} [-t_{DB}]_{00} p_D^{(0\xi_3)} \\ &\times [t_{DC}]^{(1)}_{\xi_3\xi_3'} p_C^{(\xi_3' 0)} [t_{CA}]_{00} | \psi_0 \rangle \Big\} \,.\end{aligned} \tag{7.107}$$

for the second-order term.

With our choice of pulse polarizations, we see that

$$F_A^{(0\xi)} = F_A \langle 0 | \hat{\mathbf{m}} \cdot \hat{\mathbf{y}} | \xi \rangle = \frac{m_b F_A}{\sqrt{2}} (\delta_{\bar{1}\xi} + \delta_{1\xi}) \tag{7.108}$$

(see eqns (7.48), (7.85), and (7.87)), while

$$F_B^{(0\xi)} = F_B \langle 0 | \hat{\mathbf{m}} \cdot \hat{\mathbf{x}} | \xi \rangle = \frac{m_a F_B}{\sqrt{2}} (\delta_{\bar{1}\xi} - \delta_{1\xi}) \,, \tag{7.109}$$

$$F_C^{(0\xi)} = \frac{m_a F_C}{\sqrt{2}}(\delta_{\bar{1}\xi} - \delta_{1\xi}), \tag{7.110}$$

and

$$F_D^{(0\xi)} = \frac{m_a F_D}{\sqrt{2}}(\delta_{\bar{1}\xi} - \delta_{1\xi}). \tag{7.111}$$

We need several combinations of these factors, including

$$F_A^{(0\xi)} F_B^{(\xi 0)} = \frac{m_b m_a}{2} F_A F_B (\delta_{\bar{1}\xi} - \delta_{1\xi}), \tag{7.112}$$

and

$$F_A^{(0\xi)} F_B^{(\xi' 0)} = -F_A^{(0\xi)} F_B^{(\xi 0)}, \tag{7.113}$$

as well as

$$F_D^{(0\xi)} F_C^{(\xi 0)} = \frac{m_a^2}{2} F_D F_C (\delta_{\bar{1}\xi} + \delta_{1\xi}) = \frac{m_a^2}{2} F_D F_C, \tag{7.114}$$

and

$$F_D^{(0\xi)} F_C^{(\xi' 0)} = -F_D^{(0\xi)} F_C^{(\xi 0)} = -\frac{m_a^2}{2} F_D F_C. \tag{7.115}$$

We'll make use of the eigenstates and eigenenergies of the zeroth-order Hamiltonian $H^{(0)}$ (defined above eqn (7.91)) to calculate the reduced pulse propagators and the time-evolution operators; these obey

$$H^{(0)}|\xi\rangle|N_\xi\rangle|n_\xi\rangle = \big[\epsilon_\xi + \hbar\omega(N_\xi + n_\xi + 1)\big]|\xi\rangle|N_\xi\rangle|n_\xi\rangle, \tag{7.116}$$

for $\xi = 0, \bar{1}, 1, 2$; $N_\xi = 0, 1, \ldots$; $n_\xi = 0, 1, \ldots$ with $\epsilon_0 = 0$; $\epsilon_{\bar{1}} = \epsilon + \frac{m\omega^2 d^2}{4} + J$; $\epsilon_1 = \epsilon + \frac{m\omega^2 d^2}{4} - J$; and $\epsilon_2 = 2\epsilon + \delta\epsilon$. The vibrational eigenkets in states $\bar{1}$, 1, and 2 can be obtained by spatial translation from those in the state 0:

$$|N_{\bar{1}}\rangle = |N_1\rangle = \exp\big\{-\frac{i}{\hbar}\frac{d}{\sqrt{2}}\hat{P}\big\}|N_0\rangle, \tag{7.117}$$

and

$$|N_2\rangle = \exp\big\{-\frac{i}{\hbar}\sqrt{2}\, d\, \hat{P}\big\}|N_0\rangle, \tag{7.118}$$

for a given value of N. Since the site-energy difference $\Delta\epsilon$ equals zero, all the zeroth-order q-mode potentials have their minima at $q = 0$, and

$$|n_{\bar{1}}\rangle = |n_1\rangle = |n_2\rangle = |n_0\rangle, \tag{7.119}$$

for a given n.

Since all four pulses have the same carrier frequency and the same pulse envelope, we drop the pulse index in evaluating the reduced pulse propagators; their relevant vibronic matrix elements are given by

$$\langle \bar{N}_\xi \bar{n}_\xi | p^{(\xi 0)} | N_0 n_0 \rangle = \langle \bar{N}_\xi | N_0 \rangle \delta_{\bar{n}_\xi n_0} \frac{1}{\sigma} \tilde{f}\big(\frac{\epsilon_\xi}{\hbar} + \omega(\bar{N}_\xi - N_0 - \Omega)\big) . \tag{7.120}$$

The Q-mode Franck–Condon overlaps can be calculated from

$$\langle \bar{N}_\xi | N_0 \rangle = e^{-\frac{m\omega d^2}{8\hbar}} \sum_{K=0}^{\bar{N}} \sum_{L=0}^{N} \frac{(-1)^K}{K!L!} \left(\frac{m\omega d^2}{4\hbar}\right)^{\frac{K+L}{2}} \sqrt{\frac{\bar{N}!\,N!}{(\bar{N}-K)!(N-L)!}}\, \delta_{\bar{N}-K,N-L} , \tag{7.121}$$

and the envelope $f(t) = e^{-t^2/2\sigma^2}$ has a Fourier transform $\tilde{f}(W) = \sqrt{2\pi}\,\sigma e^{-W^2/2\sigma^2}$.

Derive the Q-mode Franck–Condon overlap expression of eqn (7.121).

The final pieces are the various time-evolution operators. That for the electronic ground state is

$$\langle 0 \bar{N}_0 \bar{n}_0 | [t] | 0 N_0 n_0 \rangle = \langle \bar{N}_0 \bar{n}_0 | e^{-\frac{it}{\hbar}\left(T + E_0(Q,q)\right)} | N_0 n_0 \rangle = \delta_{\bar{N}N} \delta_{\bar{n}n} e^{-i\omega t(N+n+1)} . \tag{7.122}$$

Similarly, for $\xi = \bar{1}$ or 1, we have zeroth-order evolution operators

$$\langle \bar{N}_\xi \bar{n}_\xi | [t]^{(0)}_{\xi\xi} | N_\xi n_\xi \rangle = \delta_{\bar{N}N} \delta_{\bar{n}n} e^{-\frac{it}{\hbar}\left(\epsilon_\xi + \hbar\omega(N+n+1)\right)} . \tag{7.123}$$

The first- and second-order evolution operators in the one-exciton manifold can be obtained from a perturbative expansion with respect to the interexciton coupling,

$$[t] \cong [t]^{(0)} + [t]^{(1)} + [t]^{(2)} , \tag{7.124}$$

where

$$[t]^{(1)} = [t]^{(0)} \frac{1}{i\hbar} \int_0^t d\tau [-\tau]^{(0)} H^{(1)} [\tau]^{(0)} , \tag{7.125}$$

and

$$[t]^{(2)} = [t]^{(0)} \left(\frac{1}{i\hbar}\right)^2 \int_0^t d\tau \int_0^\tau d\bar{\tau} [-\tau]^{(0)} H^{(1)} [\tau - \bar{\tau}]^{(0)} H^{(1)} [\bar{\tau}]^{(0)} . \tag{7.126}$$

For the first-order contribution, we find vibronic matrix elements

$$\begin{aligned} \langle \bar{N}_\xi \bar{n}_\xi | [t]^{(1)}_{\xi\xi'} | N_{\xi'} n_{\xi'} \rangle = {} & \frac{\omega d}{2} \sqrt{m\hbar\omega}\, \delta_{\bar{N}N} \frac{\sqrt{\bar{n}}\, \delta_{\bar{n},n+1} + \sqrt{n}\, \delta_{\bar{n}+1,n}}{\epsilon_\xi - \epsilon_{\xi'} + \hbar\omega(\bar{n} - n)} \\ & \times \left\{ e^{-\frac{it}{\hbar}[\epsilon_{\xi'} + \hbar\omega(N+n+1)]} - e^{-\frac{it}{\hbar}[\epsilon_\xi + \hbar\omega(\bar{N}+\bar{n}+1)]} \right\} , \end{aligned} \tag{7.127}$$

which describe optical nutation; recall from eqn (7.104) the meaning of the primed excitonic indices. Because $[t]^{(1)}_{\xi\xi'}$ changes the q-mode quantum number,[16] we see that $\langle \uparrow_A \downarrow_B \uparrow_D \mid \uparrow_C \rangle^{(1)}$ of eqn (7.106) *vanishes*.

[16] This feature has the interesting consequence that in the equal site-energy case, interexciton coupling fails to generate electronic coherence between $|\bar{1}\rangle$ and $|1\rangle$ while nonetheless transferring population; see A. J. Kiessling and J. A. Cina (2020), cited on page 107.

Carry out the derivation of eqn (7.127).

It remains to evaluate the second-order evolution operator

$$\langle \bar{N}_\xi \bar{n}_\xi | [t]^{(2)}_{\xi\xi} | N_\xi n_\xi \rangle = -\frac{\delta_{\bar{N}N}}{\hbar^2} e^{-\frac{it}{\hbar}[\epsilon_\xi + \hbar\omega(\bar{N}+\bar{n}+1)]} \times \sum_{\bar{\bar{n}}} \langle \bar{n}|K(q)|\bar{\bar{n}}\rangle\langle\bar{\bar{n}}|K(q)|n\rangle \int_0^t d\tau\, e^{i\frac{\tau\nu}{\hbar}} \int_0^\tau d\bar{\tau}\, e^{i\frac{\bar{\tau}\bar{\nu}}{\hbar}}, \tag{7.128}$$

where $\nu = \epsilon_\xi - \epsilon_{\xi'} + \hbar\omega(\bar{n} - \bar{\bar{n}})$ and $\bar{\nu} = \epsilon_{\xi'} - \epsilon_\xi + \hbar\omega(\bar{\bar{n}} - n)$. A subtlety arises in handling the double integral

$$\int_0^t d\tau\, e^{i\frac{\tau\nu}{\hbar}} \int_0^\tau d\bar{\tau}\, e^{i\frac{\bar{\tau}\bar{\nu}}{\hbar}} = \frac{\hbar}{i\bar{\nu}} \int_0^t d\tau \left(e^{i\frac{\tau}{\hbar}(\nu+\bar{\nu})} - e^{i\frac{\tau\nu}{\hbar}}\right); \tag{7.129}$$

the factor $\langle \bar{n}|K(q)|\bar{\bar{n}}\rangle\langle\bar{\bar{n}}|K(q)|n\rangle$ in eqn (7.128) can be nonzero for $\bar{n} = n$, where $\bar{\nu} + \nu$ vanishes and the second integration in eqn (7.129) becomes a special case. A practical way to rescue that formula is to replace $\bar{\nu} + \nu$ by $\bar{\nu} + \nu + i\eta$, where η is numerically "infinitesimal" (we use $\eta = 10^{-8}\hbar/\tau_v$). The working expression for the second-order evolution operator is then

$$\begin{aligned} \langle \bar{N}_\xi \bar{n}_\xi | [t]^{(2)}_{\xi\xi} | N_\xi n_\xi \rangle = \hbar\omega \frac{m\omega^2 d^2}{4} \delta_{\bar{N}N} e^{-\frac{it}{\hbar}[\epsilon_\xi + \hbar\omega(\bar{N}+\bar{n}+1)]} \\ \times \sum_{\bar{\bar{n}}} \left(\sqrt{\bar{n}}\,\delta_{\bar{n},\bar{\bar{n}}+1} + \sqrt{\bar{\bar{n}}}\,\delta_{\bar{n}+1,\bar{\bar{n}}}\right)\left(\sqrt{\bar{\bar{n}}}\,\delta_{\bar{\bar{n}},n+1} + \sqrt{n}\,\delta_{\bar{\bar{n}}+1,n}\right) \\ \times \left[\frac{e^{i\frac{t}{\hbar}(\nu+\bar{\nu}+i\eta)} - 1}{\bar{\nu}(\nu+\bar{\nu}+i\eta)} - \frac{e^{i\frac{t}{\hbar}\nu} - 1}{\bar{\nu}\nu}\right]. \end{aligned} \tag{7.130}$$

Derive this expression. Based on the criterion $\hbar\omega\frac{m\omega^2 d^2}{4}\frac{t}{\hbar\bar{\nu}} < 1$, for how long can we expect second-order perturbation theory to remain valid?

Equation (7.130) completes the list of ingredient needed for numerical calculations of $\langle \uparrow_A \downarrow_B \uparrow_D \mid \uparrow_C \rangle$ through second order. Since pulses B and C are both x-polarized, exchanging B and C everywhere yields the corresponding formulas for $\langle \uparrow_A \downarrow_C \uparrow_D \mid \uparrow_B \rangle$.

We examine just the first of the two overlaps contributing to the 2D-WPI signal. Figure 7.17 shows $\mathrm{Re}\{e^{i\varphi_{BA} - i\varphi_{DC} + i\Omega(t_{DC} - t_{BA})} \langle \uparrow_A \downarrow_B \uparrow_D \mid \uparrow_C \rangle^{(0)}\}$ (divided by $m_a^3 m_b F_A F_B F_C F_D$) for a waiting time $t_{CB} = 1.50\tau_v$. The absolute value of this zeroth-order overlap is plotted in Fig. 7.18. The regularity of the peak locations in the latter plot belies considerable variation in the peak *heights*. The temporal locations of large $|\langle \uparrow_A \downarrow_B \uparrow_D \mid \uparrow_C \rangle^{(0)}|$ seen in Fig. 7.18 are governed by coherent site-to-site energy-transfer dynamics. In accord with eqns (7.97) and (7.98), peaks occur for $(t_{BA}, t_{DC}) = \frac{h}{2J}(m+\frac{1}{2}, n) = 0.378\tau_v(m+\frac{1}{2}, n)$. The size of these peaks is conditioned by the Franck–Condon dynamics of the Q-mode vibration, whose equilibrium position in both states $\bar{1}$ and 1 is displaced by $d/\sqrt{2}$ relative to that in the electronic ground state, giving rise to the same Q-mode nuclear wave-packet motion in both single-exciton states.

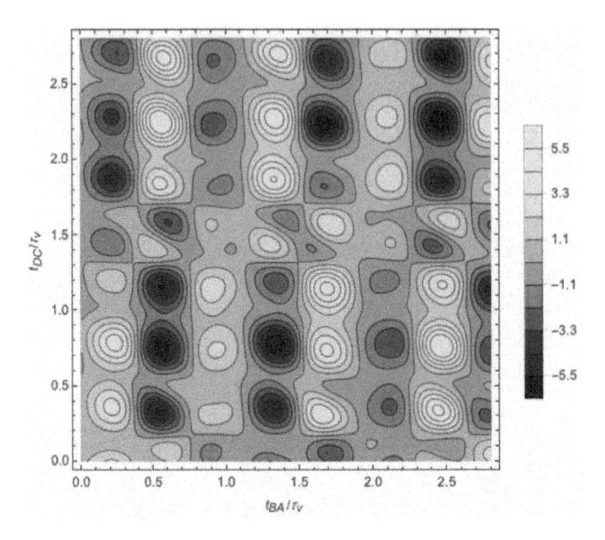

Fig. 7.17 Real part of $e^{i\varphi_{BA}-i\varphi_{DC}+i\Omega(t_{DC}-t_{BA})}\langle\uparrow_A\downarrow_B\uparrow_D \mid \uparrow_C\rangle$ in the strong EET-coupling case, calculated at zeroth order in the interexciton coupling. The pulse length is $\sigma = 0.06\tau_v$ and the carrier frequency is set to $\Omega = (\epsilon_{\bar{1}} + \epsilon_1)/2\hbar$.

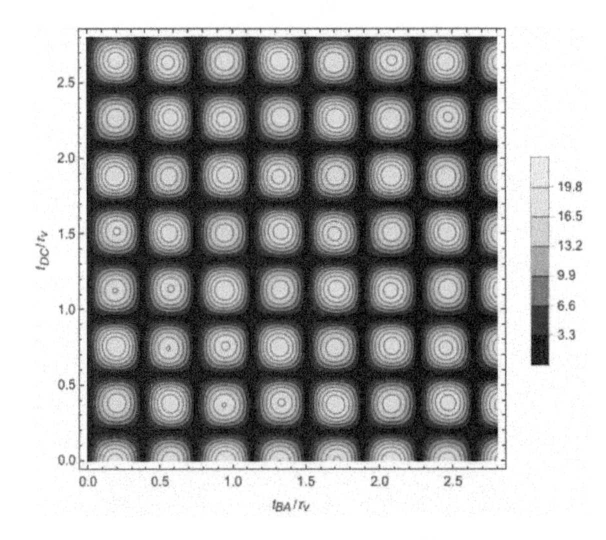

Fig. 7.18 Absolute value of $\langle\uparrow_A\downarrow_B\uparrow_D \mid \uparrow_C\rangle^{(0)}$ in the strong-J case.

For (Q, P) coincidence between the bra and ket of $|\langle\uparrow_A\downarrow_B\uparrow_D \mid \uparrow_C\rangle^{(0)}|$, the interpulse delays must obey

$$\left(1 - e^{-i\omega t_{BA}}\right)e^{-i\omega t_{CB}} + 1 = e^{i\omega t_{DC}}\,, \tag{7.131}$$

which, with the chosen t_{CB} value, leads to $t_{BA} + t_{DC} = l\tau_v$. This condition accounts for the pattern of peak intensities in Fig. 7.18.

Verify eqn (7.131) and draw phase-space diagrams for the bra and ket wave packets.

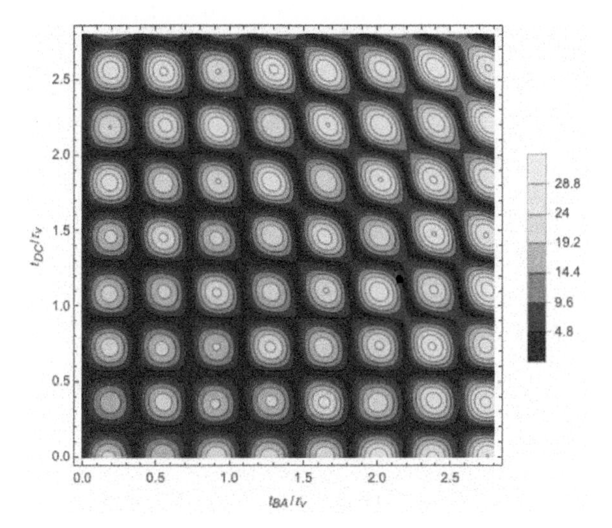

Fig. 7.19 Absolute value of $\langle \uparrow_A \downarrow_B \uparrow_D \mid \uparrow_C \rangle$ in the strong-J case calculated through second order in the interexciton coupling.

The absolute value of the same quadrilinear overlap, calculated through second order in the interexciton coupling, is shown in Fig. 7.19. Other than some subtle changes in the peak shapes, the most noticeable difference from the zeroth-order overlap is a slight shortening of the observed energy-transfer period: looking along the abscissa, for instance, we see about a quarter of an "extra" peak in two-and-a-half vibrational periods worth of t_{BA} compared with Fig. 7.18.

Although our calculations are carried out in the exciton basis, an electronically adiabatic description should nearly apply as well, since $2J$ is somewhat larger than $\hbar\omega$.[17] The singly excited adiabatic potentials of eqns (7.13) and (7.14) are found to be

$$E_{\bar{1}}(Q,q) \cong \epsilon + \frac{m\omega^2 d^2}{2} + \frac{m\omega^2}{2}\left(Q - \tfrac{d}{\sqrt{2}}\right)^2 + J + \frac{m\omega_{\bar{1}}^2}{2} q^2 , \tag{7.132}$$

and

$$E_{\bar{1}}(Q,q) \cong \epsilon + \frac{m\omega^2 d^2}{2} + \frac{m\omega^2}{2}\left(Q - \tfrac{d}{\sqrt{2}}\right)^2 - J + \frac{m\omega_1^2}{2} q^2 , \tag{7.133}$$

respectively, where

$$\omega_{\bar{1}} = \omega\sqrt{1 + \frac{m\omega^2 d^2}{2J}} \quad \text{and} \quad \omega_1 = \omega\sqrt{1 - \frac{m\omega^2 d^2}{2J}} . \tag{7.134}$$

[17]With its chosen parameter values, the model under study does not exhibit resonances between vibronic levels in its two singly excited adiabatic electronic states of the kind that can sometimes trigger nonadiabatic dynamics in energetically downhill excitation transfer. See D. M. Jonas (2018), cited on page 98.

The splitting between q-mode vibronic ground states in the two adiabatic electronic states is therefore $2J + \hbar(\omega_{\bar{1}} - \omega_1)/2$, rather than $2J$ as in the zeroth-order excitonic description. The shift of $\hbar(\omega_{\bar{1}} - \omega_1)/2 \cong h/(10.5\tau_v)$ predicts an extra electronic oscillation in ten-and-a-half vibrational periods, which is consistent with Fig. 7.19.

Appendix A
Electromagnetic energy change due to light absorption

With incidence along z and polarization along x, and switching to spherical polar coordinates, the change in energy of the electromagnetic field due to its interaction with a molecular absorber given by eqn (1.10) becomes

$$\begin{aligned}\Delta\mathcal{U} &= -\frac{E}{2\pi c^2}\int_0^\infty s\,ds\int_0^\pi \sin\theta\,d\theta\int_0^{2\pi} d\phi\, f\left(t-\tfrac{s}{c}\cos\theta\right)\cos\left[\Omega\left(t-\tfrac{s}{c}\cos\theta\right)+\varphi\right]\\ &\quad\times\left[\dddot{m}_x\left(t-\tfrac{s}{c}\right)(1-\sin^2\theta\cos^2\phi)-\dddot{m}_y\left(t-\tfrac{s}{c}\right)\sin^2\theta\cos\phi\sin\phi\right.\\ &\quad\left.-\dddot{m}_z\left(t-\tfrac{s}{c}\right)\sin\theta\cos\theta\cos\phi\right]\\ &= -\frac{E}{2c^2}\int_0^\infty s\,ds\,\dddot{m}_x\left(t-\tfrac{s}{c}\right)\int_0^\pi \sin\theta\,d\theta\, f\left(t-\tfrac{s}{c}\cos\theta\right)\\ &\quad\times\cos\left[\Omega\left(t-\tfrac{s}{c}\cos\theta\right)+\varphi\right](1+\cos^2\theta)\,. \end{aligned} \tag{A.1}$$

Show how it comes about that $\dddot{m}_y$ and $\dddot{m}_z$ do not contribute to $\Delta\mathcal{U}$.

Now,

$$\begin{aligned}&\frac{\partial}{\partial\theta}\left\{f\left(t-\tfrac{s}{c}\cos\theta\right)\sin\left[\Omega\left(t-\tfrac{s}{c}\cos\theta\right)+\varphi\right](1+\cos^2\theta)\right\}\\ &=\sin\theta\left\{\frac{s}{c}\frac{df}{dt}\sin\left[\,\right](1+\cos^2\theta)+\frac{s\Omega}{c}f\cos\left[\,\right](1+\cos^2\theta)-2f\sin\left[\,\right]\cos\theta\right\}\\ &\cong\frac{s\Omega}{c}\sin\theta f\left(t-\tfrac{s}{c}\cos\theta\right)\cos\left[\Omega\left(t-\tfrac{s}{c}\cos\theta\right)+\varphi\right](1+\cos^2\theta)\,; \end{aligned} \tag{A.2}$$

in making the last approximation, we have used $df/dt \sim \sigma^{-1}$ along with $s\Omega/c \gg s/\sigma c > 1$. Hence,

$$\begin{aligned}\int_0^\pi \sin\theta\,d\theta\, f\cos\left[\,\right](1+\cos^2\theta) &\cong \frac{2c}{s\Omega}f\left(t+\tfrac{s}{c}\right)\sin\left[\Omega\left(t+\tfrac{s}{c}\right)+\varphi\right]\\ &\quad-\frac{2c}{s\Omega}f\left(t-\tfrac{s}{c}\right)\sin\left[\Omega\left(t-\tfrac{s}{c}\right)+\varphi\right]. \end{aligned} \tag{A.3}$$

The first of these terms is negligibly small, so we have

$$\Delta\mathcal{U} = \frac{E}{c\Omega}\int_0^\infty ds\, \dddot{m}_x\big(t - \tfrac{s}{c}\big) f\big(t - \tfrac{s}{c}\big) \sin\left[\Omega\big(t - \tfrac{s}{c}\big) + \varphi\right] . \tag{A.4}$$

Letting $v = t - s/c$ and integrating twice by parts gives

$$\Delta\mathcal{U} = -E\Omega \int_{-\infty}^{\infty} dv\, m_x(v) f(v) \sin(\Omega v + \varphi) , \tag{A.5}$$

as quoted in eqn (1.13).

Appendix B

Delay regions for doubly excited-state-visiting overlaps in the difference-phased singly excited-state populations

---

We work out here the regions of $\{t_{BA}, t_{DC}, t_{CB}\}$ where the overlaps making up $P_{one}^{(d)}$ may take nonzero values. The contributing overlaps whose three-pulse bra or ket does not access the doubly excited electronic state (deriving from the first eight terms in eqn (7.39)) have the same delay ranges as the corresponding overlaps of $P^{(d)}$ for a system with a single electronic excited state, considered in Chapter 6. The admissible delays of those that do visit the doubly excited state (from the ninth through sixteenth overlaps in eqn (7.39)) are illustrated below.

$\langle \uparrow_A \uparrow_D \downarrow_B \mid \uparrow_C \rangle$ is confined to negative or very short positive waiting times t_{CB} because it depends on pulse D acting before B, so $t_{DC} + t_{CB}$ can be no larger than the pulse duration σ. The data-organizational requirement that the temporal midpoint of C and D follow that of A and B ($2t_{CB} + t_{BA} + t_{DC} > 0$) is the only other restriction, so in general this overlap can be nonnegligible in the quasi-3D region displayed in Fig. B.1.

The signal contribution from $\langle \uparrow_A \uparrow_D \downarrow_C \mid \uparrow_B \rangle$ may only be sizable in the quasi-2D delay region shown in Fig. B.2. For t_{DC} must be less than σ and, as usual, $2t_{CB} + t_{BA} + t_{DC} > 0$. The condition $t_C - t_A = t_{CB} + t_{BA} > -\sigma$ imposes no additional restriction.

The three-pulse wave packet participating in $\langle \uparrow_D \mid \uparrow_B \uparrow_C \downarrow_A \rangle$ can only be formed under the stringent conditions $t_{BA} < \sigma$, $t_{CB} > -\sigma$, and $t_{CB} + t_{BA} < \sigma$ along with $2t_{CB} + t_{BA} + t_{DC} > 0$. These requirements restrict nonnegligible signal contributions to the quasi one-dimensional t_{DC} range plotted in Fig. B.3.

Contributions to $P_{one}^{(d)}$ from $\langle \uparrow_A \mid \uparrow_B \uparrow_C \downarrow_D \rangle$ may occur only when $t_{CB} > -\sigma$. The $2t_{CB} + t_{BA} + t_{DC} > 0$ condition makes an impact only at very small intrapulse-pair delays. This important overlap may in principle exist anywhere within the 3D delay volume portrayed in Fig. B.4.

Because pulse D acts before A in its bra $\langle \uparrow_D \uparrow_A \downarrow_B \mid \uparrow_C \rangle$ can only exist for very small values of all three interpulse delays. Thus $t_D - t_A = t_{DC} + t_{CB} + t_{BA} < \sigma$ combined with $2t_{CB} + t_{BA} + t_{DC} > 0$ confines nonnegligible values of this overlap to the quasi-1D region illustrated in Fig. B.5. The overlap $\langle \uparrow_D \uparrow_A \downarrow_C \mid \uparrow_B \rangle$ is restricted to

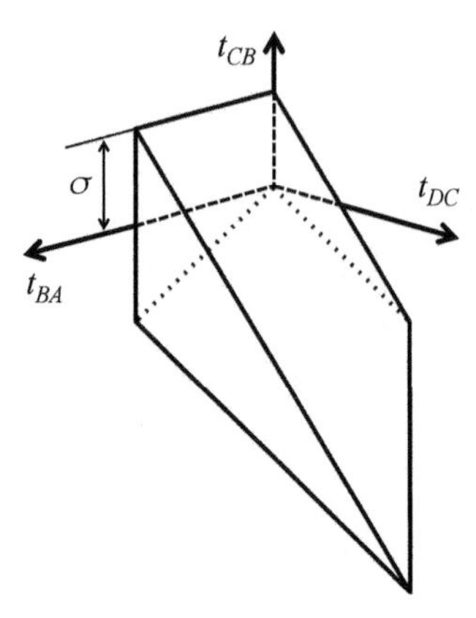

Fig. B.1 Region of interpulse delays where the effects of the order of pulse action alone do not prevent $\langle\uparrow_A\uparrow_D\downarrow_B|\uparrow_C\rangle$ from taking significant values. This delay-space volume extends to arbitrarily large positive t_{BA} and t_{DC}, and arbitrarily large negative t_{CB}. Note that this delay volume coincides with that shown in Fig. 6.2. Reprinted by permission from Springer Nature: J. A. Cina and A. J. Kiessling in *Coherent Multidimensional Spectroscopy*, edited by M. Cho (Springer Nature, Singapore, 2019).

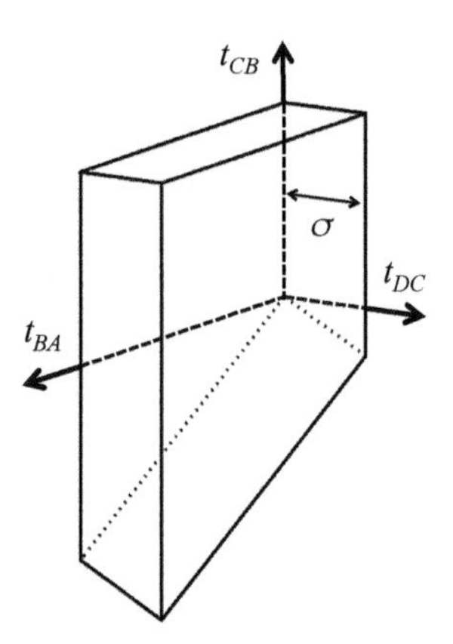

Fig. B.2 Delay region where $\langle\uparrow_A\uparrow_D\downarrow_C \mid \uparrow_B\rangle$ may be nonzero. This pulse-duration-thick slab extends to arbitrarily large positive t_{BA} and t_{CB}, and to $t_{CB} \approx -t_{BA}/2$. Reprinted by permission from Springer Nature: J. A. Cina and A. J. Kiessling in *Coherent Multidimensional Spectroscopy*, edited by M. Cho (Springer Nature, Singapore, 2019).

the same region: $t_D - t_A$ must again be less than the pulse duration, and the condition $t_C - t_A = t_{CB} + t_{BA} < \sigma$ proves to be redundant.

The ket in $\langle\uparrow_D|\uparrow_C\uparrow_B\downarrow_A\rangle$ vanishes unless both $t_C - t_A = t_{CB} + t_{BA}$ and t_{BA} alone are less than the pulse duration. Along with $2t_{CB} + t_{BA} + t_{DC} > 0$, these lead to the delay region of significance sketched in Fig. B.6.

Finally, Fig. B.7 shows the region of time-delays that may give rise to signal contributions from $\langle\uparrow_A|\uparrow_C\uparrow_B\downarrow_D\rangle$. It is largely confined to negative waiting times, because

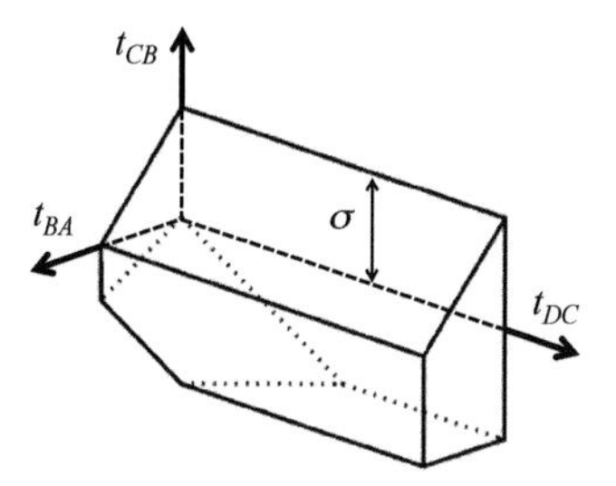

Fig. B.3 Narrow region of delay-space to which nonnegligible contributions from $\langle\uparrow_D|\uparrow_B\uparrow_C\downarrow_A\rangle$ are confined. Reprinted by permission from Springer Nature: J. A. Cina and A. J. Kiessling in *Coherent Multidimensional Spectroscopy*, edited by M. Cho (Springer Nature, Singapore, 2019).

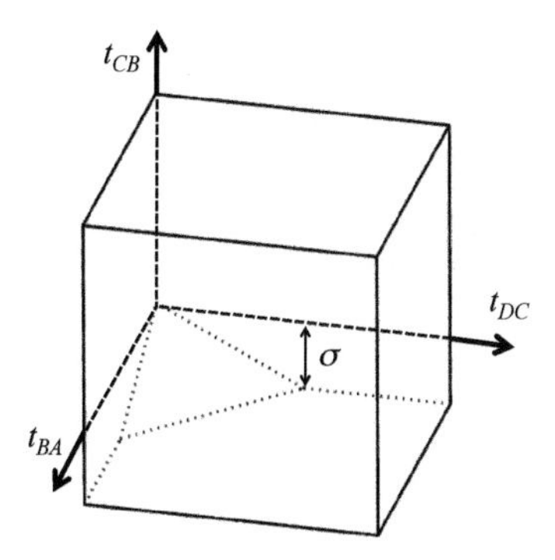

Fig. B.4 Volume in $\{t_{BA}, t_{DC}, t_{CB}\}$, extending to indefinitely large values of all three delays, where $\langle\uparrow_A|\uparrow_B\uparrow_C\downarrow_D\rangle$ can in principle make a signal contribution. Reprinted by permission from Springer Nature: J. A. Cina and A. J. Kiessling in *Coherent Multidimensional Spectroscopy*, edited by M. Cho (Springer Nature, Singapore, 2019).

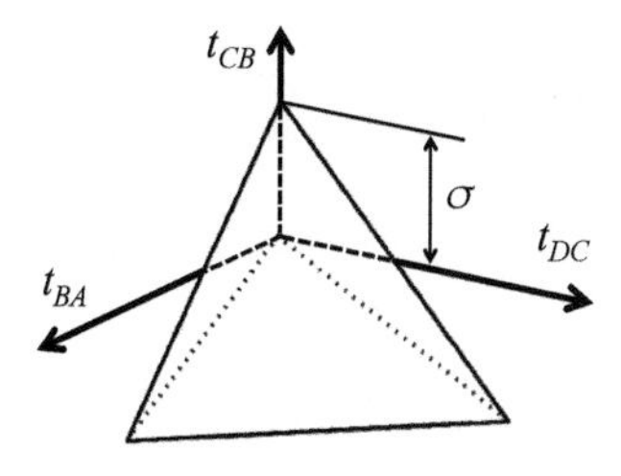

Fig. B.5 The small delay-range region within which both $\langle\uparrow_D\uparrow_A\downarrow_B|\uparrow_C\rangle$ and $\langle\uparrow_D\uparrow_A\downarrow_C|\uparrow_B\rangle$ are expected to contribute to $P^{(d)}_{one}$. Reprinted by permission from Springer Nature: J. A. Cina and A. J. Kiessling in *Coherent Multidimensional Spectroscopy*, edited by M. Cho (Springer Nature, Singapore, 2019).

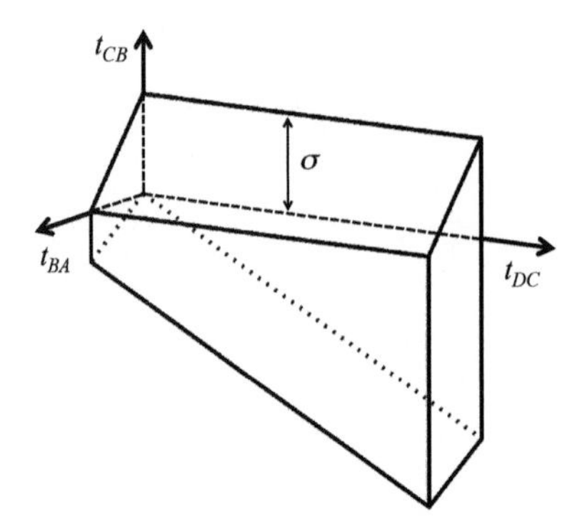

Fig. B.6 Interpulse-delay region, extending to arbitrarily large positive t_{DC} and negative t_{CB}, where $\langle \uparrow_D \,|\, \uparrow_C \uparrow_B \downarrow_A \rangle$ may give rise to significant signal. Reprinted by permission from Springer Nature: J. A. Cina and A. J. Kiessling in *Coherent Multidimensional Spectroscopy*, edited by M. Cho (Springer Nature, Singapore, 2019).

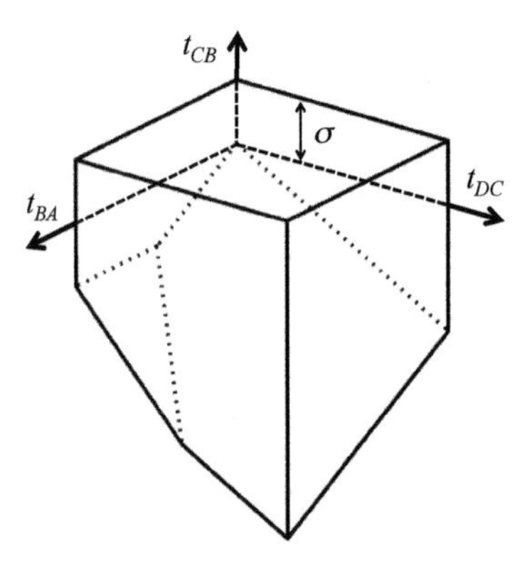

Fig. B.7 Quasi three-dimensional delay volume, extending to indefinitely large positive t_{BA} and t_{DC} and to arbitrarily large negative t_{CB}, within which the overlap $\langle \uparrow_A \,|\, \uparrow_C \uparrow_B \downarrow_D \rangle$ can contribute to the signal.

t_{CB} must be less than the pulse length. Since B acts before D in the three-pulse ket, $t_D - t_B = t_{DC} + t_{CB} > -\sigma$, and, as usual, $2t_{CB} + t_{BA} + t_{DC} > 0$.[1]

[1]The delay volume for this overlap was inadvertently omitted from Cina and Kiessling (2019), cited on page 96.

Appendix C

Delay regions for overlaps contributing to the difference-phased doubly excited-state population

In this appendix, we work out the region of $\{t_{BA}, t_{DC}, t_{CB}\}$ to which each of the difference-phased quadrilinear contributions to the population of the doubly excited state, seen in eqn (7.41), are exclusively confined, while taking rough account of the nonzero duration $\sim\sigma$ of the ultrashort laser pulses. In $\langle\uparrow_A\uparrow_D|\uparrow_B\uparrow_C\rangle$, the pulses act in their nominal order (A before D and B before C), so the overlap can be nonnegligible in a three-dimensional region of delay-space. More specifically, t_C must exceed $t_B-\sigma$, or the B-pulse could not act before C; hence $t_{CB}>-\sigma$. But the bookkeeping restriction $t_D+t_C>t_B+t_A$ (see Chapter 6) also requires $t_{CB}>-\frac{1}{2}(t_{DC}+t_{BA})$, which is slightly more restrictive for tiny intrapulse-pair delays. All three interpulses delays can take arbitrarily large positive values. These are the same restrictions as apply to $\langle\uparrow_A|\uparrow_B\uparrow_C\downarrow_D\rangle$, so the relevant delay volume is identical to Fig. B.4.

In the ket of $\langle\uparrow_A\uparrow_D|\uparrow_C\uparrow_B\rangle$, C-pulse action precedes B-pulse action. In order that this overlap not vanish, t_{CB} must therefore be shorter than σ. As the general condition $t_{CB}>-\frac{1}{2}(t_{DC}+t_{BA})$ also applies, nonvanishing values can only reside in the delay region plotted in Fig. C.1.

The overlap $\langle\uparrow_D\uparrow_A|\uparrow_B\uparrow_C\rangle$ can only be nonnegligible when $t_D-t_A=t_{DC}+t_{CB}+t_{BA}$ is less than the pulse duration. But t_{CB} must exceed $-\frac{1}{2}(t_{BA}+t_{DC})$, so this contribution to $P_{two}^{(d)}$ can exist only in a quasi zero-dimensional region of very short interpulse delays. The same restrictions apply to $\langle\uparrow_D\uparrow_A|\uparrow_C\uparrow_B\rangle$. For nonnegligibility of this latter overlap again requires $t_{DC}+t_{CB}+t_{BA}<\sigma$. The condition $t_{CB}<\sigma$ is weaker and adds no additional constraint. The resulting common delay region coincides with that illustrated in Fig. B.5, which confines $\langle\uparrow_D\uparrow_A\downarrow_B|\uparrow_C\rangle$ and $\langle\uparrow_D\uparrow_A\downarrow_C|\uparrow_B\rangle$.

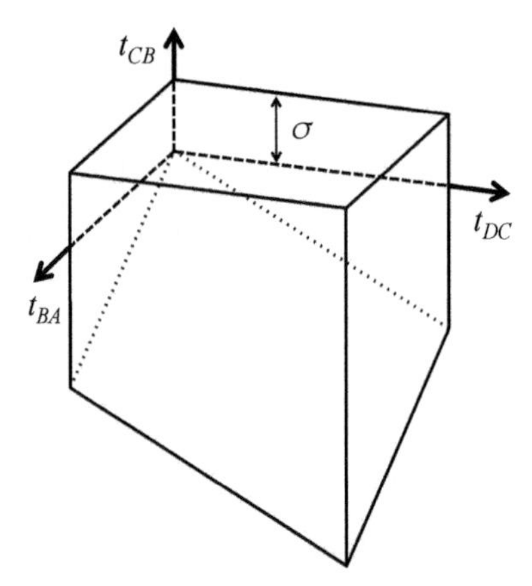

Fig. C.1 Region of interpulse delays where $\langle \uparrow_A \uparrow_D \mid \uparrow_C \uparrow_B \rangle$ can be nonzero. Reprinted by permission from Springer Nature: J. A. Cina and A. J. Kiessling in *Coherent Multidimensional Spectroscopy*, edited by M. Cho (Springer Nature, Singapore, 2019).

Index

absorption, excited-state, 21, 22, 55, 56, 58, 59, 109, 114, 118
absorption, linear, 2, 5–8, 17, 23–26, 94, 129
absorption, transient, 17–19, 22, 23, 25, 53–55, 75
adiabatic approximation, 11, 39–41, 46, 52, 53, 55
adiabatic electronic states, 96, 98, 100, 104, 127, 128
adiabatic theorem, 11, 41, 46, 52, 53, 55
adiabatic vector potential, 14, 15
adiabaticity, electronic, 11, 13–15, 39, 40, 45, 96, 97
adiabaticity, vibrational, 35, 46, 47, 53, 55, 56, 59
Albrecht, Allison W., 62
Almy, J., 87
Anna, Jessica M., 105
Apkarian, V. A., 87
Arpin, Paul C., 60
Aspuru-Guzik, Alán, vii

Baym, Gordon, 11
Berry, Michael V., 15
Biggs, Jason D., 50
Born–Oppenheimer approximation, 11, 15
Brazard, J., 33
Brixner, Tobias, 69

Cerullo, Giulio, 53
Chatterjee, Sambarta, 107
Cho, Minhaeng, vii, 62, 96
Cohen-Tannoudji, C., 90
Condon approximation, 1
conformational coordinate, 23, 24, 26, 34, 35, 44, 46–48, 52, 53, 55, 56, 58, 59
cumulant expansion, 87, 89, 90, 93

density matrix, viii
density operator, viii, 6
Dhar, Lisa, 30
Dietze, D. W., 34
dipolar field, radiated, 2
dipole moment, induced, for electronic absorption, 2
dipole moment, induced, for four-wave mixing, 69–71, 76, 82–84
dipole, induced, for femtosecond stimulated Raman spectroscopy, 36–38, 45, 47, 49, 50
dipole, induced, for transient absorption, 18–20, 22, 54

electronic decoherence, 102
electronic excitation transfer, 96–99, 105–110, 112, 115, 118, 120, 121, 125, 127
energy transfer, electronic, 96–99, 105–110, 112, 115, 118, 120, 121, 125, 127
energy transfer, strong-coupling case, 99, 118–128
energy transfer, weak-coupling case, 99, 105–118
exciton basis, 98–100, 119–121, 125
exciton shift, 111–119

femtosecond stimulated Raman spectroscopy, FSRS, fissors, 34, 36, 44, 45, 47–51, 53, 57, 59
Fleming, Graham R., vii, 62, 105
four-wave mixing, FWM, 61, 69–76, 82–84, 99
Franck–Condon displacement, 32, 35, 45, 55, 57, 58, 81, 90, 94, 108
Franck–Condon overlap, 5, 8, 23, 24, 28, 31, 45, 57, 58, 124
Franck–Condon region, 25, 49, 50
fun, isn't this, 93

Gaynor, J. D., 62
Gera, R., 105
Ginsberg, Naomi S., 105
Glauber coherent states, 16, 90
ground-state bleach, 4, 21, 22, 27, 29, 31, 56, 64, 65, 81, 121

Hamm, Peter, 62
Harris, Robert A., vii
Heller absorption overlap, 5, 8, 24, 25
Heller, Eric J., vii, 5
Hochstrasser, Robin M., 62
Huang, L., 105
Humble, Travis S., 107

impulsive stimulated Raman excitation, 21, 22, 27, 29–34, 56, 121
interaction picture, 3, 41, 77, 103
interexciton coupling, 120–122, 124–127

Jang, Seogjoo J., 98
Jonas, David M., 62, 69, 98, 127

Jumper, Chanelle C., 18, 53, 60

Khalil, Munira, 62
Kiessling, Alexis J., 96, 107, 124, 134
Kilin, Dmitri, 107
Kovac, Philip A., 18, 60
Kuramochi, Hikaru, 34

Lee, S. Y., 34
local oscillator, 62, 69–74, 76, 84
Longuet-Higgins, H. C., 16

Makri, Nancy, 107
Marcus, Andrew H., 69
Mathies, Richard A., 33, 34
McCamant, David W., 34, 49
Moruzzi, V. L., 16
Mueller, S., 69
Mukamel, Shaul, vii, 50, 62, 89

Nelson, Keith A., 30

Ogilvie, Jennifer P., 105
optical dephasing time, 30, 94, 118
optical phase locking, phase shift, 61–64, 68–70, 72, 76, 96, 101, 102, 117
optical response function, viii, ix, 89

partition function, 7
Pauli operators, 16, 41, 96, 97
perturbation theory, time-dependent, 3, 4, 20, 21, 41, 54, 76, 78, 88, 103, 104, 110, 122, 124, 125
perturbation theory, time-independent, 14, 39
phase signature, sum or difference, 64–66, 68, 70, 101–103
phase-space plot, 80, 81, 90–92, 112–116, 127
Pollard, W. T., 33
Polli, Dario, 53
pulse duration, 5, 6, 8–10, 17, 20, 21, 23, 27, 28, 31, 36, 47, 50, 53, 56, 60, 63, 65, 67, 72, 73, 75, 78, 79, 85, 104, 105, 111, 120, 121, 131, 132, 135
pulse propagator, reduced, 4, 21, 22, 27–30, 42, 45, 55–57, 60, 77–80, 83–86, 98, 104, 109–111, 121, 123, 124

quantum yield, 101–103, 109, 111, 117, 118

Raman gain, 43, 49, 50
Raman shift, 34, 48, 49
Rice, Stuart A., vii, 62
rotating-frame transformation, 38, 42, 43
rotating-wave approximation, 4, 5, 21, 38, 78, 84, 104

Sakurai, J. J., 14
Scherer, Norbert F., vii, 62
Schlau-Cohen, Gabriela S., 105
Scholes, Gregory D., 18, 53, 60
semi-classical Franck–Condon approximation, 41, 55, 56
Silbey, Robert J., vii
single-molecule spectroscopy, 69, 79, 101, 105
Slonczewski, J. C., 16
Smith, T. J., 30
Song, Y., 105
spatial-translation operator, 57, 123
spectral filtration, 18, 27, 37, 38, 54, 73–76
spectral interferometry, 61, 73–76
stimulated emission, 21, 22, 27, 28, 31, 32, 56, 64, 81–84, 87, 88, 92, 121

Tahara, Tahei, 34
Tanimura, Yoshitaka, 62
Tannor, David J., vii
Tekavec, Patrick F., 69
thermal equilibrium, 6
Tiede, David M., 105
time-circuit diagram, 87, 88
Turner, Daniel B., 33, 60

vibrational decoherence, 49, 50
vibrational dephasing, 49
vibrational frequency tracking, by fissors, 52
vibrational spectroscopy, multidimensional, 62

wave-packet interferometry, WPI, whoopee, 61, 62, 69, 79, 84, 85, 96, 99, 107–109, 111, 114, 116–119, 125
wave-packet overlap, ix, 5, 8, 24, 25, 37, 45, 47, 56, 57, 63, 64, 68, 79, 81, 86, 87, 90, 101–106, 111, 117, 121, 125, 131, 135
wave-vector matching, 72, 79, 82–84, 101
Wong, Cathy Y., 61

Yuen-Zhou, Joel, vii

Zanni, Martin T., 62